Reading Science:

A Literary, Political, and Sociological Analysis

THE REYNOLDS SERIES IN SOCIOLOGY

Larry T. Reynolds, *Editor*

by **GENERAL HALL, INC.**

Reading Science:
A Literary, Political, and Sociological Analysis

Ben Agger
State University of New York
Buffalo

General Hall, Inc.
Publishers
5 Talon Way
Dix Hills, New York 11746

Reading Science:
A Literary, Political, and Sociological Analysis

GENERAL HALL, INC.
5 Talon Way
Dix Hills, New York 11746

Publisher: Ravi Mehra
Consulting Editor: Anand Sinha
Composition: *Graphics Division,* General Hall, Inc.

LIBRARY OF CONGRESS CATALOG CARD NUMBER: **89-80379**

ISBN: 0-930390-93-8 [cloth]

Manufactured in the United States of America

For John O'Neill
whose 1969 Hegel-Marx seminar
got me started

Science narrates. Yet its positivist version does everything possible to conceal its literary nature. Facts are just facts; science writers merely report the world; speculative reason falls victim to methodology. Science presents a chilly world in which inequality flourishes precisely to freeze it. As such, it accelerates the falling rate of intelligence. But not only can we read—and oppose—science as literary and political discourse; science can essay itself in a different way, neither distancing itself from literature nor renouncing objective knowledge. Here I read science in order to change the world it pretends only to reflect. Science constructs.

Contents

Acknowledgments

I would like to thank the following people for offering extraordinarily constructive suggestions about the manuscript: Linda Brodkey, Norman Denzin, Paul Diesing, Paula England, Joe Feagin, Tim Luke, John O'Neill, Allan Rachlin, Beth Anne Shelton. I am grateful to Larry T. Reynolds for sponsoring the book in his series with General Hall. Ravi Mehra, President of General Hall, made production a breeze. Irene Glynn did a fine job editing the book. Lionel S. Lewis, my chairperson in sociology at SUNY-Buffalo, unfailingly offers support and encouragement.

Ben Agger
Buffalo, New York
February, 1989

Chapter 1 Reading Writing Differently

Academic discourse increasingly suffocates thought.[1] That is common knowledge today among many of us who experience academia directly and confront the academic discourse congealed in journals and monographs. Yet asserting it does not really get to the heart of the issue; knowing it is one thing, making the critique available for an audience is something else. In particular, having a critical method at hand with which to penetrate the obscure discourses of the academy would both allow one to translate obscurantist formulations into accessible language and offer a rule of writing that avoided the pedantry and pretense of self-mystifying expert argot. It is not enough to know that academic discourse stifles critical imagination. It is also important to read in a way that disgorges the thought buried in it; after all, even falsehood educates.

Marxism, for many years, has suggested a rule of demystification sometimes called the critique of ideology. Where ideology is false consciousness, an inaccurate version of the surrounding world, its critique penetrates illusion and thereby liberates.[2] In falsifying false versions, we gain access to the truth deftly suppressed by falsehood in the first place. While the Marxian critique of ideology has been revised quite fundamentally since the Russian Revolution and World War II, it still models a method of interpretation with which to confront the cultural deceptions of a social order founded on exploitation. Without worrying about the pedigree of ideology critique, indeed without even requiring obedience to the hallowed concept of ideology, I want to extend ideology critique to the texts of social science. In this book, I read some examples of mainstream social science writing, particularly sociology (my discipline), as the literary—and thus political—texts they are. Demystifying their false representations of an invariant social world, I suggest a critical method with which to address other cultural expressions.[3] I want to read journal science as an ironically self-concealing literary strategy.

Here I essay science as narrative, not as a vouchsafed method guaranteeing an undistorted reflection of the world. Unlike fiction, science methodically claims validity for itself; Shakespeare does not append a section on his database or research procedures but simply assumes that we read *The Tempest* as the truthfully intended account it is. Sociology must convince its readership of its fidelity to reality, showing its epistemological credentials through the citation of authority and the display of methodology. In the process, a positivist version of science conceals its literary nature as narrative in order to shape a world it would reflect as invariant. Positivist science surreptitiously provokes behavior along the lines of adjustment and adaptation it implies are fixed.

Suppressing its rhetorical nature, positivism acts politically where it conceals its own literary contribution to the construction of the world. It does so precisely to conceal the mutability of the world; by suppressing its narrative nature, its fictive quality, science's text ironically proves the truth of the world it pretends to reflect presuppositionlessly. Positivist social science postures a world it knows is incomplete in order to complete it. If people live out the frozen world depicted in science, then science's announcement of that world turns out to have been true all along. If such things as poverty, racism, and sexism are inevitable, then people act irrationally when they undo them. Science politically conceals its own political nature.

Having said that as a way of beginning, I spend this book demonstrating how science narrates a false world out of readers—that is true for having been enacted as such. In the chapters that follow I rewrite journal science narratively. I read the text of mainstream social science as a literary accomplishment supported by certain rhetorical, technical, and figural devices constituting the dense text of the journal article. In this, I do not mainly appraise the "findings" of science as it announces them in the section of the article reserved for that report. Instead, I "find" the findings in all sorts of unexpected places—in the margins, at the bottom and top of the page, in the notes and bibliography, figurally in the text, usually everywhere but in the words themselves (although sometimes there, too).

Science narratively advances its case for a frozen world that its account would only reinforce by suppressing narrative. It does this through configurations of text and space and rhetorical rhythms that seduce the reader into the inexorable sense of an immutable social horizon. Let me offer an initial example. The most prestigious American sociology journal, *American Sociological Review* (here-in-after *ASR*), ran an article recently on racial variations in the age of first intercourse for young people.[4] It is not enough to note only the banality of the topic or the authors' interpretations of their findings; even more important is the way in which the argument is crafted compellingly *to perpetuate itself* and thus the world sheltering it. Articles of this kind littering mainstream science journals reproduce the political and social world indirectly by adding value to the canons or "literatures" of mainstream disciplinary knowledge proceeding narratively in the same way. Thus, such an article provokes certain behavior out of individuals directly; indirectly, it reproduces a discipline legitimated by authority and methodology. By appearing in print, the article both adds value to others before it and provokes ones in its image. Science reproduces itself (and thus the world it freezes representationally) apart from its announced research findings.

Within science's dominant writing style, certain things happen. Virtually all articles suggest or imply causality in the social world even if those words are not used. Studies of the variation of first intercourse by race empower themselves to disclose a seamless world in which volition plays little role in the outcome of events. Science writes the way it does in order to suggest invariant causal patterns or at least probabilistic connections not to be undone by defiants. Science is conservative not merely because its practitioners advocate a certain explicit social agenda or because its announced findings lead to conservative policies but because its text pretends to reflect a frozen world—concealing its own literary contribution to the deep freeze.

At issue is the narrative status of methodology.[5] While most social science disciplines inculcate students with research methodology often involving a heavy dose of statistical analysis, these are really courses in narrative style. Methodology narrates a narrationless world, thus reproducing it. If successful, methodical science appears purely to reflect the world without presuppositions; the muddying presence of the author is expunged from the text, the text is buttressed by its affiliation to discipline, and the data are displayed in the midst of the text as the world's sheer presence. Whereas social science students are routinely taught that methodology is one's value-free mode of approach to research topics and data, ensuring objectivity and the ability to generalize, I reconstruct methodology as the ability to write science in a way that makes it convincing as an ontological account of social being. Science's literary method is political practice wherein the world under account is frozen in the lens of description. Positivism is a literary strategy that aims to expunge authorial trace from the published text and thus to ensure a reading that enacts the text's account of the world as an immutable world itself.

By now, the observation that positivism seeks invariant social laws has become second nature to critics of it. I agree with that observation. But few who write everyday science worry about lawfulness; in fact, advanced methodologists caution agnostically against the unjustified inference of causality from messy data probed with relatively primitive techniques. Instead, normal science speaks the language of correlations, employing the complex statistical argots littering the journals with code, figure, equation, table.[6] Social science's positivism is conveyed narratively, from the first page of journal articles through the endnotes and bibliography.[7] It is easy for everyday scientists to eschew positivism as either old-fashioned or controversial, thus saving them many ontological headaches. Positivism is less a doctrine than a literary style, a way of constructing an incontrovertible world out of textual fragments—data, words, acknowledgments, citations.[8]

Methodology narratively squeezes narrative out of the text. As the possibility of different formulations of itself, history stands in the way of the text's postured seamlessness. Thus it is expunged through the deceptive ontology of methodology.

Others have done content analyses of mainstream sociology journals, showing how trivial are the topics considered by authors as well as demonstrating some common disciplinary themes.[9] I presuppose this work and do not want to repeat it. While content is important, I search for it less in the words themselves than between, underneath, and around them. This book is about literary method particularly as science's rhetorical strategy binds it to a vouchsafed disciplinary canon and postures a presuppositionless and ahistorical reflection of the world at hand. I want to see how science *narrates* on the page.[10] By so doing, I can better understand the deft sleight of hand worked by those who write for and control the mainstream journals in constructing a common sociological world. I oppose that world simply because positivism remains the enemy of history and thus of radicalism. People are not ineluctably patterned, caused, governed, or correlated; indeed, this recognition empowers them to resist. Yet while data can be highly liberating when they show people the structural hinges of their oppression, the positivist display of data suggests them as invariant features of a mathematically and figurally reflected world.[11]

Although I find that science uses number and figure as its main vehicles of expression, there is nothing inherently quantitative about positivist writing. Mainstream journals sometimes publish "qualitative," even theoretical, articles. These articles may have different topics and titles, but they too proceed in the fashion of quantitative science: They are subordinated to canon and made to reflect a frozen world. As such, although mainstream social science relies heavily on the languageless narratives of quantity, figure, and symbol to portray a historyless world, nonquantitative approaches often proceed in much the same way without relying explicitly on techniques of statistical inference. At issue is not quantification per se but the way that text harnesses figure to speak for it, thus establishing its epistemological authority.

Science persuades,[12] whereas the pages of many journals and monographs appear hard and fast, merely reflecting an intractable world. It is one thing for polemicists to write essays about the eternity and ubiquity of economic inequality. It is another thing for people to encode those arguments methodologically in mainstream journals, where words and data feed off one another in an unspoken symbiosis of deception and overinterpretation. In the first instance, polemic invites response. In the passion of its own presentation, polemic essays a position that inherently could be different. It practically invites challenges; indeed, it is usually written in counterpoint to something else. Thus Rousseau wrote "A Discourse on the

Origin of Inequality" in a bold attempt at speculative reason. But the two sociologists Davis and Moore precluded dialogue in their *ASR* article "Some Principles of Stratification,"[13] in which they dispassionately announce the inevitability of economic inequality (even though, by today's standards, their argument seems speculatively unsubstantiated by the accouterments of methodology).

Although this book is about the literary style of journal science, I also suggest a different approach to sociological writing. Critique suggests construction as I move through the argument. For example, in chapter 3 I discuss the figural and textual arrangement of journal articles' typical opening page. In particular, I examine the author's intended connection to discipline, authorities, and topic designedly figured in the page's format. In looking for text when text is absent, I suggest not only a way of reading for subtle and deceptive meaning but also a way of writing foregrounding one's narrative intent. It is the peculiar literary characteristic of positivist science that it disqualifies authorial self-consciousness out of hand.[14] Not only first-person prose is illegitimate but a host of other features of what one might call essay—a counterpoint to the science article. I do not end this book with a recipe for "good" writing, but I do suggest some implications of my critique of science for different writing in particular and for emancipation in general.

Although I seem to work within the interpretive frame of reference of literary theory in my "deconstruction" of scientific writing, my approach to literary theory joins it to social theory. I search for a larger world hidden in the interstices of the positivist text; I want to learn from falsehood. In particular, I want to theorize the suppressions as well as embellishments of the positivist text as a way of understanding cultural suppressions and embellishments generally. Self-consciously risking overdetermination in saying this, *every account has significant meaning for the world it would describe or conceal in its spiritless acount.* Nothing about science's presentation is accidental, and nothing can be ignored. Literary theory offers insights with which to read texts for their undercurrents, even unconscious. But literary theory without a critical theory of society guiding it fails to understand the relationship between text and world. Derrida is wrong: The world is not only a text, although text itself is a nucleic social relationship both produced and reproduced in larger layers of social structure.[15]

When I read science's discourse, I try to understand its literary strategy ironically as saying both more and less than was intended. All versions are political in the relationship they strike to the larger world. In pretending only to reflect a seamless totality, positivism (knows it) thereby reproduces it, provoking behavior out of "readers" convinced of the ahistorical truthfulness of their depicted happiness, reasonableness, even apathy. Thus I politicize what seem to be merely conventions of literary style. While I disagree with Derrida that everything is a text, the text is certainly a world of its own in which power is transacted between and among readers and writers. To use Marxist language, I want to explore science as a text concealing and revealing certain social relationships of knowledge, notably the hierarchy between credentialed insiders and marginalized outsiders.[16] Of central importance here is my discussion of the way science bends over backward to ground itself canonically, citing dominant authorities, acknowledging powerful friends, and working within the vouchsafed research methodology—only another name for the art of persuasion. The question of literary style becomes intensely political when style imitates, patronizes, genuflects to the established tradition.

It is increasingly apparent that American sociology is dominated by a midwestern quantitative empiricism emanating out of Michigan, Wisconsin, Ohio State, and other land-grant institutions. Ivy League sociology departments, especially Harvard, have disdained quantitative research as crass, productivist, and overly mechanical, preferring a more speculative "qualitative" mode of scholarly presentation. While times may be changing in the Ivy League (witness the tenure denial of Paul Starr by Harvard after he won the Pulitzer Prize for a nonquantitative book accessible to a nonspecialist audience—followed by his hiring at Princeton), the major journals and monograph publishers are dominated by a narrative approach to methodology rooted in fetishistic citation and figural procedures. One can understand this on a number of levels. First, a sociologist might simply do a network analysis of the shift of disciplinary power from Chicago and Harvard to public midwestern and sunbelt institutions. Or one might understand the rise of quantitative method as myth in a late stage of capitalism steered by certain technical elites. These are both true as far as they go. In this book I want to engage midwestern empiricism by looking at the dominant sociology journals exemplifying a methodological approach to everyday science.

Readers unfamiliar with developments in European literary theory, especially the shift from structuralism (Saussure, Levi-Strauss, Barthes) to poststructuralism (Barthes, Foucault, Derrida), may regard literary style as a nonissue, especially when everyone knows that the leading journals all speak in the same deauthorized technical prose. For these people, style refers to irrelevant issues of presentation, or, worse, style may be heard to characterize precious forms of discourse such as fiction and poetry. Since the rise of aesthetic theory bearing on social and political questions, both within Marxism (Lukacs, the Frankfurt School, Goldmann) and outside it (poststructuralism), issues of style are directly pertinent to the analysis of authorial meaning and social structure. Although I do not want to make the thicket surrounding aesthetic theory today even denser, my reading of scientific journal presentation borrows heavily from contemporary theories of interpretation. Together these approaches to reading, to textual interpretation, enrich Marxist and feminist perspectives on what Gramsci first called cultural hegemony—ideology, in Marx's classic terms.

It is a central tenet of Western Marxism, first in Lukacs and then the original Frankfurt School, that science has become a form of domination in late capitalism. Although certain orthodox Marxists misread critical theory to suggest that instrumental rationality (positivism, high technology, computers, etc.) has displaced the logic of capital as the moving force of history, a variety of heterodox Marxists from Lukacs and Korsch to Horkheimer and Adorno argue that positivism is more than a form of knowledge; it is also a way of life inuring people to their alleged fates in late capitalism. As ideology, positivism flattens the historical universe of wide-open possibilities into the "one-dimensional" appearances of the present.[17] Positivism as what Wittgenstein called a language game[18] binds people to their supposed fates by involving them in a passive relation to the object of knowledge—here late capitalism. The positivist pursuit of invariant social laws, the desideratum of Comte's "social physics," takes a literary form susceptible to cultural interpretation and criticism.

As the Frankfurt School has aptly demonstrated, cultural criticism of this sort need not proceed without reference to social, political, and economic injustice.[19] In order to resist the reductive determinism of orthodox Marxist aesthetic and cultural theory—Zhdanovism—we need not subordinate political economy to the realm of ideation, spirit, culture. As Horkheimer demonstrated over a half century ago (1972), a dialectical theory of society moves back and forth between economy and culture, refusing any such distinction as categorially arbitrary and empirically fruitless. In much the same vein, I regard "textuality" as crucial in late capitalism, not because I seek the truth in text, as many would, but because texts, books, and articles are a moment, a factor, in domination. While scientific writing within particular academic disciplines may count for very little in the larger scheme of things, it represents a general kind of disciplining in which people are habituated to their exploitation through their participation in language games disempowering them cognitively and constitutively—positivism.[20] Positivism is a political factor, nothing more or less; it accelerates the falling rate of intelligence.[21]

It is far-fetched to suggest that the *ASR, American Journal of Sociology,* and *Social Forces* more than marginally abet capitalism. But to the extent to which sociologists who read these journals take a positivist literary strategy into the classroom and employ it in their own writing, sociological positivism is not to be ignored. And to the extent to which Comte's social physics is a modal positivism, it is worth reconstructing the literary, thus political, logic of science's text. Literary theory meets politics where interpretation addresses the obfuscation, as well as upsurge, of meaning in texts. When Marx identified religion as ideology in early market capitalism, he did not pause to deconstruct its texts; it was sufficient in a secularizing Europe to denounce religion as obviously false. But Marx could scarcely afford the same luxury when he confronted the ideological tomes of bourgeois political economy. The volumes of *Capital* are exercises in textual criticism, unraveling the contradictoriness of political economy as a way of showing their ironic truths about the money economy of emerging capitalism.[22]

Positivism is the new version of ideologizing discourse, enticing submission with its postured lawfulness. Sociology as an exemplar of positivism does not replace capitalist economic theory as disempowering myth but supplements it with institutional accounts of the eternity and necessity of our bondage to Durkheim's social facts and Weber's iron cage—the power of the social. Quantitative sociology was prefigured early in the Enlightenment as social arithmetic; Comte only gave disciplinary voice to this literary project. The *ASR* can be read in much the way Marx read Smith and Ricardo. In concealing the open possibility of history in fealty to canons and in its figural and statistical charades, mainstream sociology as a literary strategy portrays a world that does not exist. As such, it wants to produce the very world it announces as a fissionless totality. This is the function of every ideology. In presenting the world ontologically it ensures that the ontology of the moment will prevail. People locked in are led to throw away the key.

It is conventional wisdom in the philosophy of social science that positivism is dead.[23] True, if positivism is found only in the rarefied air of philosophy departments. But as a literary strategy it governs virtually all mainstream writing in sociology and such neighboring disciplines as political science and economics. The number crunchers in the Midwest do not read Hempel and Ayer; they have not paused over Popper and his debate with Adorno. The Vienna Circle could as well describe a group of Freudians. Yet, *as writers,* the midwesterners are positivist—that is, they suppress the trace of their own constitutive authoriality in their literary work. In pretending to portray the world presuppositionlessly in the text, *they thus aim to bring it about.* The establishment of motive is irrelevant; it does not matter that the quantitative empiricists understand their own subtle contribution to the totality. Indeed, few do. That is precisely the nature of routine science.[24] They only "do it," plowing ahead in their descriptive work heedless of the larger epistemological, and thus political, auspices of it. They disdain what they call "theory" either as the hagiography of disciplinary founders or simply as museum work, tending the corpses of the dead.[25] Above all, midwestern empiricists are impatient; they want to get on with the business of building a discipline, entrenching the canon of citations and authority ladders comprising the narrative institution of journal science.[26]

Sociologists would understand conventionalism conventionally.[27] That is, they would describe the emergence of literary usages in terms of the internalization of literary norms governing professional apprenticeship. This is not wrong as far as it goes. Literary style is imparted in graduate school, if not earlier. In order to publish before they earn their Ph.D.s, thus increasing their marketability as budding academics, graduate students must learn how to write "journal articles," presenting research in the standard format accepted and expected by the main journals in the discipline.[28] Term papers are crafted into this format, with attention paid to norms governing style—citation, bibliography construction, acknowledgment of "literature" to which the paper's topic relates, description and analysis of data, the "discussion/conclusion" section.[29]

While convention governs literary usage in mainstream science, this does not say enough. It does not understand what appears to be merely literary technique in larger political terms. Thus, the citation of authorities in acknowledgments, notes, and bibliography bears political implications for the pedigree of the argument. The display of data in graph and figure is more than a technical choice; it implies a conception of the world made accessible through number and design. Literary technique discloses an ontology, a theory of social being, that its repetition only reproduces. Positivist technique promotes a positivist conception of the world, and thus a positivist world. The conventionalist explanation of literary usage indulges in circularity. The interpretive task, instead, is to understand what appear to be purely technical and textual issues in terms of a a larger social and political world to which they bear an indirect relation. Conventions describe what happens when power becomes a norm. The authorial choices taught in graduate school are not random; they reinforce a certain conception of the world, and thus the world itself.

Let me offer an example, to be pursued later in the text. Most academic journals print one's institutional affiliation underneath one's name at the top of the first page of one's published article. Many journals elaborate one's reputation at the front or back of the particular issue of the journal in which one's work appears. This elaboration might be called "about the contributors." While it seems obvious that people's institutional affiliations would be printed under their names on page 1 of their articles, *this assumes that they have institutional affiliations*—that they submit the article under letterhead and can be addressed at their institutional addresses. The *ASR* in a footnote on the first page, further indicates the institutional mailing address of the author to whom questions and criticism may be addressed by interested readers.

People who do not have institutional affiliations are less-than-full disciplinary citizens.[30] Occasionally one may see an author addressed as a "private scholar," but this is extremely rare in mainstream journals. By contrast, the radical journal of social theory, *Telos,* no longer prints institutional affiliation, presumably because many people working in theory, especially Continental critical theory, are unemployed or underemployed. The busy midwestern empiricist will probably find this discussion trivial in the extreme, hardly disqualifying the published "results." But what appears to be merely a technical convention speaks powerfully about what is to be regarded as legitimate discourse. Certainly, people who submit articles without letterhead (not to mention from prestigious institutions) have less chance of acceptance by journals and university presses. Thus the naming of people's institutional affiliations only deepens the equation of legitimate, worthy scholarship with regular university employment.

Moreover, while many journals list one's institutional affiliation, they do not differentiate that by department. Thus contributors to the *ASR* are assumed to be sociologists, just as contributors to the *American Political Science Review* are assumed to be political scientists. Sociologists publish sociology in sociology journals. And they know where to send their work; some journals are sociology journals and others are not. Furthermore, most professionalized sociologists who pay attention to their discipline understand that *ASR, AJS* and *Social Forces* are the "big three," with other disciplinary journals trailing behind in importance. Although authors are assumed to be regularly employed as sociologists, they also know where to send their work for maximum impact and thus career advancement.

It is tempting to regard the inclusion of institutional affiliation as a mere folkway, a trivial social trend changing with the times. But styles do not originate de novo; they are constructed, learned, internalized—*narrated.* They tell us something about the world they intend to reproduce. Some sociologists, following Parsons, would say that these folkways have a function; whether or not one would use that language (I would not, for function implies "good" practice), the printing of institutional affiliation underneath an author's name has signifying power. It reinforces the division between insiders and outsiders. It also signifies an institutional pluralism where people from lesser-known universities and sociology departments, even graduate students, publish in major journals alongside established people from famous institutions. The folkway bears meaning; it does not happen for nothing. But saying that does not imply an original conspiracy. Significance takes shape in history. As disciplines institutionalize, these small signs of differentiation become hierarchies. Publishing in "major" journals is better than publishing in "lesser" ones; by the same token, having an institutional affiliation is better than being a sole scholar, just as being from a reputable institution is better than being from obscurity.

This is the kind of technical-textural analysis I present here. I read journal science for its echoes, even its unconscious. I refuse to take it simply on its own terms, nor do I relativize its literary style as mere convention.

Something else is going on. Science subtly signifies its superiority to non-science in order to reproduce a world in its image. Although we have to dig beneath appearances to "hear" this message, it is there, unmysteriously. But literary interpretation usually addresses established literary canons or even revisionist canons as long as the cultural artifacts in question are clearly "fiction."[31] Few read science for its literariness, let alone the political sense of its rhetorical style. Science is wrongly read only at its surface, missing the imaginative depth out of which style emerges as an important text in its own right. When many professional sociologists read the major journals, they are overwhelmed with difference; the articles are heard to contribute narrowly to the discipline's subfields, innovating in small ways. When I read professional sociology, I am struck by the unrelieved sameness of the texts in question. Thus I can reconstruct them as a single account of the social world, more continuous than discontinuous.[32]

Employed sociologists sometimes complain that the major journals exclude certain styles of intellectual work. That is truer than they usually know. Such critics are usually near the discipline's methodological and theoretical center, but not quite close enough to have their work accepted. They clamor for more pluralism in editorial policies. Indeed, two Marxist authors are actually grieving formally against *ASR* for rejecting their "Marxist" work out of hand. Although I am also an outsider, I do not clamor to be inside. Differences melt down in the enmeshing text of science. No matter what, deviation is softened under the logic of science's narrative style. This is not to say that I would refuse publication if it were offered. That is beside the point. But I could not be published without having my text put through the wringer of domestication. The argument would be boiled down into "findings," mere reportage of a literature review (as it is called) and "discussion." A recent *ASR* editor is known to ask authors of "theory papers" to "propositionalize" their articles, forcing them into the hypothetical mold of empirical science.

The major sociology journals refuse to publish diverse theoretical and methodological styles not at the center of disciplinary discourse—critical theory, feminist scholarship, phenomenology. Occasionally works using these words see the light of day, if only as caricatures, but the real problem is that illegitimate work is disqualified as nonsociological, not simply as the theoretical school in sociology it is supposed to represent. The only article ever to treat critical theory published in *ASR* or *AJS* attacked critical theory as subversive of intellectual and political freedoms, virtually as Stalinist. The author, a graduate student at a Canadian university, was allowed to have the first and last word.[33]

Thus one will find the occasional reference to Marx and feminism in mainstream journals. But one will not find nondisciplinary, even antidisciplinary, work—work that does not respect the boundaries of the discipline as outlined by the founding positivist-functionalists—Comte, Durkheim, and Weber. To get published in sociology journals one must write sociology, work clearly inhabiting the discipline's implicit parameters of canon and method. One must cite the same sources other people do, and one must duplicate their research methodologies, predominantly quantitative empiricism. Most social science disciplines by now are highly specialized, canonized, and methodologized; they have their stars, their literatures, their appropriate techniques.[34] It is not specialization per se that disqualifies heterodox work; nor is the solution simply "interdisciplinary" work, whatever that might mean. The issue, rather, is the degree of hyper-canonization and methodologization of particular disciplines *as literary styles.*[35] To "be" a sociologist one must write as sociologists write, imitating their references, overarching theoretical frameworks, and methodologies. In this book I want to unpack the literary style of methodology in its own right—literally, a method of presentation.[36] To get published as a sociologist in sociological journals one must learn and replicate this style, including both rhetorical elements (mode of presentation) and appropriate obeisance to dominant literatures or canons (social relations of presentation).

The mode and social relations of contemporary sociology are dominated by certain ways of beginning articles (chap. 3), canonizing topics in relation to an established literature (chap. 4), writing and figuring findings (chaps. 5–7), and concluding (chap. 8). These are my main topics as I read recent examples of articles published in leading American sociology journals. I do not offer precise content analyses but provide symptomatic readings supporting my reconstruction of the mode and relations of scientific presentation. In chapter 9 I discuss the "theoretical" article and its place in journal science. In chapter 10 I conclude with a discussion of different approaches to the reconstruction of science's narrative, recommending essay as an improvement over methodology.

I do not cull articles scientifically from carefully sampled journals. To do so assumes that science's own methods can be used uncritically in examining it.[37] I work within the dominant journals in order to understand them simultaneously from the inside and outside. In effect, they can be rewritten as literary fictions, manipulating words, space, and figure in order to create a luminous world on the page. As a theory of knowledge and discourse, positivism attempts to capture the world essentially in text and word, without the aid of mediating presuppositions or dialectically interrogatory talk. This book rewrites science as it was written and thus as it could be read. I do not accuse it of dissembling but read its suppressions and marginal embellishments as well as its supposed content. Content is inextricable from context where texts are social material in their own right. Books and journals matter because they are matter; they play a political role in culture simply by existing, available for readings that turn into lives.[38]

This is not to burden disciplinary sociology with too much. Certainly neither the journals nor sociologists directly aid the contras, threaten nuclear holocaust, send dissidents to Siberia, assault women, or cause poverty. Nor do journals and sociologists simply invert the "real" world in deceptive—ideological—prose constructions. Yet science in its literary work reproduces a conception of the world, and thus a world itself, governed by intractable laws, thus diminishing the emancipatory imagination in times that need it most. Positivist sociology models and deepens a positivist world; alternatively, sociology could model liberating language games and mount intellectual challenges to the present world. It does not because the literary conventions of positivism are in fact politically grounded and driven. Even if the hard-working midwestern empiricist may not recognize the political auspices of his or her work, they are there, captured in the most mundane elements of narration in academic journals. Positivism perpetuates a world in which people cede knowledge to experts, and experts subdivide that world into analytically separable elements joined in relations of causality.[39]

In this book I want to make science read as the literary, hence political, text it is. While not poetry, mainstream science as reflected in the dominant sociology journals is no less fiction than Vonnegut or Graham Greene. But unlike fiction, science's narrative style reproduces itself, provoking the rate of intelligence to fall further. It is tempting to demonize positivism in the fashion of many left theories whose concepts get out of hand, run wild, and become thoughtless slogans. My study of journal science convinces me that positivists model an intractable world (that they secretly know they must buttress with their own fatalizing version of it); they write incredibly trivial articles dominated by methodological and figural pyrotechnics, thus ensuring the eclipse of reason. I disassemble the text of journal science in showing how it proceeds from its opening page through footnotes and bibliography. I do that here in the interests of demystifying and then reformulating a hegemonic positivism. Academic sociology is certainly not the only discipline dominated by quantitative empiricism with its numbing, numbering mindlessness.[40] Lives are being wasted writing and publishing what is little more than gibberish; positivist texts both reinforce the world they reflect ontologically and prevent people from asking more fundamental questions. Positivism is a literary style subordinated to canons that would reproduce subordination as a rule of contemporary social organization; it is also an enormous waste of time. We could be doing other things, prefiguring other worlds.

Chapter 2 Scripting Causality

Mainstream sociology is positivist, even if it does not say so. It is precisely because it does not. Instead, workaday empiricists, in filling the journals with arcane statistical, figural, and descriptive analyses, reinforce the dominant conception of the world and preclude other narrations of it. Journal science in this regard portrays a lawful social world in which fundamental political defiance is disqualified by the data themselves—the imperviousness of the late-capitalist world to fundamental challenges. I want to propose a literary and political understanding of science's text that largely dispenses with labored discussion about the intellectual pedigree of positivism; indeed, the argument that positivism is passe strengthens it. Meantime, positivist writers continue to spin their literary fancies on the pages of mainstream academic journals. While I largely restrict my discussion to American sociology, I could cite a number of articles from an "interdisciplinary" social science journal, showing that the tendency is general. Positivism narrates itself in a standard way. I want to read it as a script of causality, thus "causing" a lawful social world fundamentally closed to political interventions.

The politics of science lie less in its uses or its explicit political self-consciousness than in its literary strategy—how it sits on the page and thus enters readers' minds and lives.[1] After all, positivism is distinct from other theories of knowledge in that it allows itself no guiding self-awareness, no literary indulgences or metatheory. Positivism is practiced, not thought.[2] To understand it we must unravel its generic text combining the rampant use of number and figure and the slavish subordination to a literature narrowly constraining topic choice. Taken together, the use of figure and number to present data and the reliance on established authority to frame the problem produce a lawful account of the social world. In the figural language of this strategy, let me outline my reconstruction of science's text and then discuss its implications for the closer readings I offer in the following chapters. This choice of figure itself should indicate that there is nothing inherently wrong with figure, number, or even the citation of authority. At best, these literary accouterments focus an argument where they are written and read to encode their self-consciousness as narrative artifacts, rhetorical choices that could have been made differently. I return to the possibility of a different version of science in my concluding chapter.

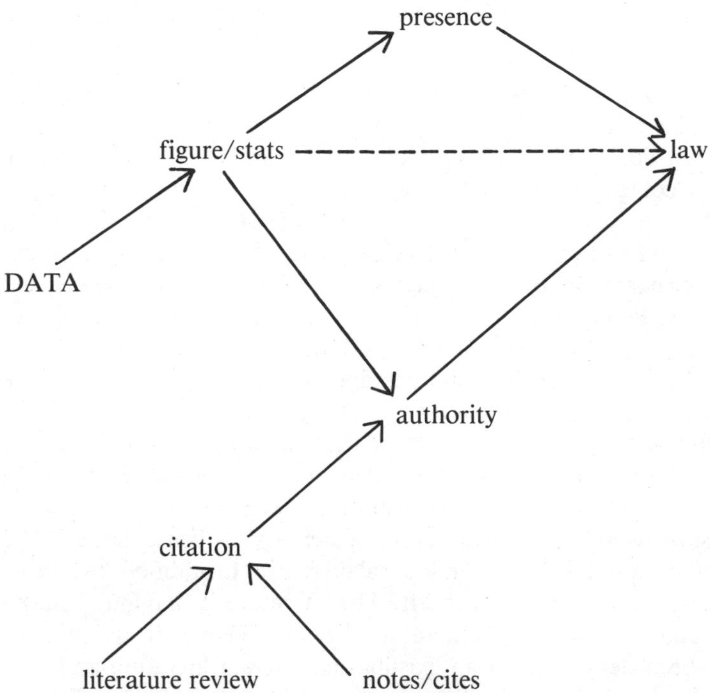

Figure 2.1

In my study of mainstream sociological journal articles I find a recurrent pattern both in quantitative and qualitative articles (see figure 2.1). Data enter the literary apparatus of the text in figural, especially numerical, form. As well, the paper's topic is contextualized within a larger literature or canon giving credence to the problem and specializing it. The use of figure operates indirectly to create the impression of the data's patterning into invariant social laws by suggesting the world itself in the pages of the text. The use of statistical analyses operates directly to enhance the appearance of the data's lawfulness, notably through the operations of statistical inference. The citation of authorities in the beginning of most articles both authorizes the analysis in terms of pregiven research parameters, lending credence to the paper's findings as indubitable and thus lawful, and narrowly contextualizes the paper's topic.[3] In addition, when the location of an analysis in a preexisting disciplinary literature directly establishes credibility, the use of figure and number indirectly enhances the argument's authority. And the establishment of an analysis' authoritative context proceeds through an initial literature review linking an analysis to earlier analyses yet defining its specificity. Authority also establishes itself through the citation of sources in notes and an extensive bibliography at the end.

Science supports its portrayal of a lawful social world through the figural display of tables, charts, and equations, as well as in the use of inferential statistics. While the inscription of figure and number supports the text's silent contention that it has captured the object world representationally on the page, figure and number support the authority of an analysis. Figure and number as rhetorical devices have mystique; as such, they enhance narration with the appearance of heightened validity, even exotic expertise. People who criticize the fetishistic tendency of quantitative analysis often conflate the figural importance of design, equation, and number in legitimating an analysis intending to represent the world with the authorizing importance of quantitative method. They are different. Figure and number both represent the world and lend authority to the analysis they help support. Both are important in understanding positivist discourse. In order to support the illusion that the analysis achieves or at least approaches lawfulness, the world must appear in transparent form—hence the elaborate use of design, equations, tables in journal literature. In addition, the use of math deepens the argument's authority when its authority has initially been established in the context of a literature review. Where a trivially narrow topic has been derived from an existing mainstream literature and thus authorized, it only gains legitimacy when it is handled figurally and mathematically.

Taken together, an analysis must both represent the world directly and derive locally from the authority of a canon for its account of lawfulness to be convincing. Thus far, I have mentioned the ways in which figural and mathematical analysis both represent the unmediated object world, the central aim of positivism, and deepen the account's authority in an age of number. There is an additional advantage to be derived from explicitly statistical analysis. When, as I intend to show, the use of figure and number takes a variety of forms, many journal articles in emulating the natural sciences directly employ the apparatus of statistical inference with which to generalize from samples to whole populations, indeed to all historical possibility.[4] When statistics are employed in an analysis, they directly enhance its apparent lawfulness. They also indirectly enhance the aura of lawfulness by both suggesting the world's presence on the page and establishing the authority of the analysis. But statistical analysis belongs to a larger class of figural and representational gestures by the text. As such, it is only one among a host of rhetorical devices for moving the prose forward. Indeed, one of my central topics below is the way analyses are figured in mainstream journals to accomplish a variety of ends including representation, legitimation, and inference.

Of course, such a model of science's literary strategy abstracts too much; fine-grained differences exist in the actual texts that do not find their way into such a reading. But a schematic overview is possible precisely because mainstream science is so homogeneous in its logic of presentation; indeed, in generalizing about journal discourse one could move way beyond sociology, even the social sciences, to natural science, the original model of positivist narrative. It has been said before that during the Industrial Revolution social sciences imitated such natural sciences as physics in seeking a lawful understanding of the social universe.[5] My claim is that this understanding is a purposeful design of artifice, an accomplishment in prose and figure—a veritable literary outcome. Laws do not exist; indeed, they cannot be approached through the gradual accumulations of what Merton called middle-range science. Sociology's agnosticism about causality is disingenuous. Workaday science claims little more than correlational knowledge but nonetheless hopes to achieve finality through piecemeal insight. Unfortunately, the parts do not—cannot—add up to the whole; unlike nature, humanity is free. That is precisely why the literary effort of positivism is so politically disabling. Trivial analyses proliferate; we deceive ourselves and our readers that the technical and figural mumbo jumbo in the meantime bring the world closer to the end of history.

Figure and Number as Ontology

Let me further elaborate on the model proposed above before I enter into closer textual analysis. Figure, number, data analysis, the citation of authority are legitimate literary strategies in other political contexts; that is, there is nothing inherent about figure, number, hypothesis, the citation of authority and data analysis that compel bad sciences—sciences purposely inattentive to their own buried political motive, namely to reflect, thus reproduce, the given social order. In this book's final chapter, I reflect on alternative literary strategies that do not abandon empirical analysis in the light of what I have said before.

The figuring of data in science's text accomplishes two direct aims; when one explicitly uses statistical inference, one accomplishes a third aim.[6] Too many critics of mathematization fail to note what really happens in the scientific text. Many seemingly mathematical approaches are not mathematical at all but merely figural. For example, causal models or equations do not necessarily entail the logic of statistical inference; frequently, articles do not present any data at all but simply suggest new ways of figuring old data or thinking about causal relationships. Although the use of statistics belongs to the general process of figuring the text, it does not exhaust figure. Precisely because it does not, one can find ample examples of nonstatistical analyses, even "theoretical" ones, that rely heavily on figure but not at all on mathematical inference.

The first "text" of figure is its inscription of the world directly in the pages themselves. The use of design and number is intended directly to reflect or represent the world at large. Figure is a window on the world allowing it to enter the text without proseful mediations. Of course, the presuppositionless reflection of the world is a central feature of positivism. The world is figured in hope of drawing the reader into the illusion of the world's literary presence. I am not saying that this is good or bad in itself; indeed, I drew my own figure a few pages earlier as a way of accelerating this discussion. But in the positivist text, figure does the crucial work of suppressing narrative. As such, then, the appearance of the world's textual presence enhances the impression that the analysis points toward lawful social patterns. The use of figure suggests an indubitability to the analysis not enjoyed by prose alone.

The second "text" of figure is the authorization of one's analysis. In and of itself the use of figure and number increases the apparent validity—believability—of one's writing; mathematics has mystique in this stage of Western history.[7] Indeed, it is one of the deepest faiths of our age. It is scarcely imaginable that science could proceed without the elaborate use of figure and number. A colleague of mine once remarked about a book I had written that it lacked "tables," weakening its epistemological strength. He meant that tables, in and of themselves, conveyed the impression of scientificity. Of course, he was not wrong, only uncritical. The authority of figure is indisputable; that is why even nondatabased articles frequently use pages of figure to boost their text to the heights of science. If number cannot be entered directly into the text, then at least one can conjure up some nifty graphs, charts, even formal models (equations) suggesting further databased research into the subject at hand.[8]

Science relies heavily on its math effect. Most working scientists believe that mathematics does important intellectual work in data analysis.[9] But, like methodology generally, statistical inference is only a poetry, a rhetoric suggesting possible patterns of relationships between and among phenomena. It is a literary device chosen to make arguments convincingly. As such, the use of math and figure has become a powerful norm in most social science disciplines trying to catch up to the natural sciences in institutional importance. Sociologists who work with numbers and formal models are regarded at the cutting edge of the discipline. Those who do qualitative research (interviews, ethnography, archival work, theory) enjoy lower status. Indeed, one of the most rapidly growing trends in sociology is "theory construction," the buttressing of theoretical work with testable propositions and even formal mathematical models.[10] Theory construction knows well how powerful is the math effect; by embellishing turgid text with figural design, the status of the thought is raised. Careers are made out of this stuff.

While the figuring of data has the twin effects of suggesting the world's pretheoretical presence in the text and authorizing one's analysis through the math effect, statistical inference as a special case of the use of figure suggests time-invariant patterns of causality, or at least approximates them. Inferential statistics, in allowing one to extrapolate data collected from samples to entire populations, directly mathematizes causality. Most sociology articles use social statistics as a narrative of causality, even if almost everyone foreswears definitiveness in the light of the less-than-perfect correlations reported in the journals. The point is not to achieve perfect causalities but merely to inch closer to them; thus the incredibly complicated mathematical elaborations now employed in social science are thought to move us cumulatively toward causal understandings.

As such, when analyses are mathematized, they not only suggest the world's presence and acquire a higher authority as cutting-edge science; they also point directly to the formulation of social laws. Virtually everyone who does databased research, even people who work with relatively small samples, tries to render their work statistical. However rigorously authors may deny the pretense of causality, the very rhetorical apparatus of statistical inference implies approximated causal patterns. So-called regression analysis, the stock in trade of contemporary mathematical social science, charts the one-way or reciprocal variations of discrete variables expressed in correlation coefficients. The figuring of these coefficients takes up huge amounts of space in journals and monographs; they represent the traces of causality, the craftwork of scientists trying to capture the elusive truth through narrow statistical routines.[11]

It is simply false that statistics do not imply causality; they imply it even if they rarely achieve it (so-called perfect correlations, where two variables unaffected by other "intervening" variables "vary" together in all cases, thus establishing what one can presume is causality). Now it is crucial here to distinguish between the use of statistical inference as a causal strategy and its use as a literary strategy. The two are different. Virtually everyone who uses statistics in social science believes that they are essential to science and that science aims toward lawful understanding, allowing one to predict.[12] Every research methodology textbooks says this. Virtually everyone who uses the statistical apparatus believes it. Math is the essence of science; science seeks laws. But statistics can be used narratively—*indeed, they are so used today.* Statistics suggest patterns and thus afford insights; they do not decide arguments definitively. Yet few people who use them acknowledge this. For most, statistics are an essential part of the scientific apparatus, getting us closer to ultimate causality. In this book I read methodology as essay—merely another venture in speculative reason.

Statistics, like figure generally, are used rhetorically to strengthen, not settle, arguments. No amount of technical manipulation will make the data speak;[13] the data are only constructions, epigrammatic pieces of history, not things themselves. But most journal scientists do not understand or admit this. Math is thought to resolve arguments definitively, and thus much space is taken up with pyrotechnical displays of figure and number. Indeed, sociology articles are less about the world represented in data than about the technical methods used to probe it. If these analyses advance our knowledge, it is largely methodological advance.

Citation, Authority, Validity

An authoritative analysis suggests lawful patterns. This explains why the empirical articles in mainstream journals inevitably begin with discussions grounding the particular topic in a larger "literature" or canon and end with elaborate citation of sources, references, authors. Such buttressing takes place in the text where names of referenced authors appear in parentheses (the favored style of citation) and even sometimes under the text where footnotes are printed at the bottom of the page instead of as endnotes. When the use of figure and number helps establish the world's presence in the text and redoubles the text's authority in its own right, the analysis also depends on its grounding in an authoritative literature making way for it.[14]

This canon and its corresponding authority are established in two ways. First, the author cites other authorities in endnotes, footnotes, bibliography and the text itself.[15] Journal articles abound with names, the dates of their publications, and sometimes the page numbers of their texts. In itself, this citation apparatus creates the impression of the article's authority. It is unimaginable that one would submit an article to a journal without citations, notes, and an extensive bibliography even if one had read widely in those sources. Second, the article is grounded in a "literature review," usually opening the paper. This section of prose utilizes many of these names and dates in showing that the author has read other writers on the same subject. In doing this, the author crucially grounds his or her analysis in a preexisting literature on the subject (thus reproducing the validity of the literature as legitimate repository of definitive knowledge) and derives a differentiated topic from the literature that will occupy his or her analysis in the article at hand. Scholars on race relations are cited as authorities in order to contextualize a particular contribution to the race-relations literature.[16]

Thus the citation of literature involves a certain irony in the literary text of science. On the one hand, one must give one's work pedigree by deriving it from the authoritative work of others. On the other hand, one must differentiate one's own work significantly so that one can acquire license to write and publish. If one is successful in this, one adds to the impression of the authority of one's analysis, especially if one employs quantitative methods. In the larger sense, one's analysis will be read to approach lawfulness if one figures and numbers the results (showing the world's presence and adding authority to the argument) and if one grounds the analysis in an apparatus of citation establishing the disciplinary relevance of one's topic.

Work within a narrow corner of an established field adds value both to the dominant literature and to one's small contribution to it. The more value is added, the more the analysis will be seen to contribute to the overall growth of knowledge. It is striking how internally referential the mainstream articles in fact are. Within subfields, one will find very little deviation from the citation of certain established sources. Citations reproduce themselves. This is easy to understand in terms of a principle of literary repetition. Each additional article takes its bearing from the articles before it. They provide the context authorizing one's own small contribution to the overall consensus. Even if one takes issue with established authorities, one only overturns those particular authorities, not the authority of the research literature itself. In fact, most articles do not overturn much of anything; they conclude inconclusively by suggesting that little has been learned beyond some rudimentary insights, thus justifying the call for additional research—the standard exit line of science articles.

The more one adds value to an existing literature, the more one can claim validity, approaching lawfulness, for one's own analysis. Even if in fact one contradicts basic findings in the cited literature, simply naming it adds value to one's own version. It is instructive that few citations in science articles include the precise page numbers of the work noted; instead, one will usually find strings of parenthetical names and their dates of publication, separated by semicolons and placed within parentheses: (Smith 1979; Smith and Jones 1983a, 1983b; Agree 1985; Disagree 1986). These string citations compound authorities without referring to specific formulations in their work. Thus the author takes substantial literary license in compressing others' views into support for his or her own analysis. Authority moves on when it is named but not really analyzed. While one's own work benefits, the subfield or literature within which one labors also gains value.

Authority is enhanced by both the apparatus of citation, within and outside the text itself, and by the math effect derived from the figuring and numbering of one's analysis. When figure is not present, one must work extra hard to establish the validity of one's analysis in elaborate citation strings as well as in an extensive bibliography. It is interesting to note a trend in recent years away from elaborate discussion of cited authorities, even in notes. The more notes diminish, the longer bibliographies become in turn. Subtly this moves science away from careful cultivation of literatures and toward the numbing effects of serial citations, simply compounding authorial names into an unargued aggregate of established knowledge.

Apart from the issue of authority and validity, one cannot even conceptualize topics without the aid of literature reviews. Whereas such reviews suggest overall coherence in one's work, they also contextualize it, connecting it to the work of others but also differentiating it sufficiently to afford one room to move. Thus one can understand the trivialization of science topics in terms of the hermeneutic requirement that articles be derived from dominant literatures, both to give one's work value (and, indirectly, adding value to the literature cited) and to specialize it in view of existing problematics of research and theory. One learns in graduate school how to make contact with established knowledge in finding one's own niche within it, thereby expanding it. After all, science in authorizing itself with respect to given authority also expands that authority in a piecemeal way, what Kuhn called normal science. Everyday science, the kind published in journals, does not shake the paradigmatic foundations of subdisciplines, let alone disciplines, but usually innovates in tiny, trivial ways, notably in terms of methodological refinements. Thus the trivialization of everyday science can be understood less in terms of authorial failure of nerve than as a requirement of discipline itself. If established literatures are basically sound, then one can contribute to them only iteratively.

Editors of mainstream journals would dispute this account, claiming that they publish both path-breaking work and narrowly innovative analyses. As always, this is an empirical question. Are the paradigms actually shifted in the journal literature? My canvass of mainstream sociology journals suggests to me that there is little innovation and that what innovation occurs is largely technical in nature. Discussion of methodological technique having little to do with the substance of the analyses at hand occupies an extraordinary amount of space in the journals. The relevance of one's work to larger intellectual issues is given short shrift; indeed, as I explore later, the "discussion" of results is usually reserved for the last page of the article—quite incredibly. The bulk of the paper is taken up with literature review, methodological description, and discussion and presentation of findings. While purely technical and descriptive articles used to be published as "research notes" at the end of the journal numbers, today the line between real articles and research notes has faded. If anything distinguishes research notes from articles, it is their somewhat shorter length, perhaps reflecting the editorial decision that the work they present is relatively unimportant.

Triviality is built into the canonical nature of journal science; one must take own's cue from established work in the field, citing it to authorize one's own analysis. Triviality is also guaranteed by the fragmented nature of the field itself. There is no single "literature" in sociology but a host of subliteratures – family, race, urban, poverty, gender, theory, and so on. In fact, these are not distinctive subfields either but huge umbrellas in their own right; they are further broken down into narrow units dominated by a handful of influential scholars who define the relevant problems for legitimate investigation. The article on age of first coitus by race that I mentioned earlier is a case in point. Narrowness reflects and thus reproduces the extraordinary degree of disciplinary specialization and differentiation. People cultivate their own small corners of the field, following the lead of their mentors in graduate school and colleagues with whom they publish.

Of course, this is a circular logic. The more differentiated discipline becomes, the more difficult it is to publish without reducing one's focus narrowly enough to support oneself with reference to an extant body of expertise. In sociology one simply cannot tackle large issues such as class and gender inequality without zooming down to the theoretically, empirically, and methodologically differentiated traditions in which these larger issues are studied. If one tries to remain at the nearly global level, anonymous manuscript referees disqualify the article on grounds that it fails to work within, or cite, narrow lines of scholarship already established in the field. And editors doubt the credibility of analyses aiming too high, dismissing them as empirically unsupported essays. Thus one must be highly specialized at the outset; in turn, the more one specializes in response to disciplinary requirements, the more one adds to that specialization and differentiation.

Scripting Causality: The Literary Politics of Science

Taken together, the figuring and numbering of analyses along with the contextualization of analyses in established scholarship contributes to the overall impression of the objective validity of an analysis.[17] In particular, this validity serves the impression that the social world is governed by laws each of which is progressively approximated by the narrowing strip-mining taking place at the nether levels of the discipline. While neither authors nor editors summatively aggregate past articles into a comprehensive version of the deterministic whole, the intention is clear: The work we are doing adds to work before it; eventually we will arrive. Until then, we must keep busy doing the local work of testing and developing subliteratures out of which larger insights can one day be drawn.

Nobody says this, even though all believe it. Everybody has read the methodology textbooks in graduate school, perhaps even delved into the positivist philosophy of science. The social sciences, particularly sociology, exist to ape the natural sciences; one day, with enough methodological rigor and accumulated research, they will join biology, chemistry, and physics in the pantheon of science. The social universe will be understood in terms of its invariant patterns and thus mastered, at least predicted with certainty. My claim here is that the pursuit of social laws is carried out in writing employing both figure and citation as the mode and relations of intellectual production, respectively. Positivism is a literary strategy, a way of presenting an inert, lawful world and not a theory of knowledge explicitly claimed or utilized by practitioners.[18] As such, positivism is fiction designed to provoke certain readerly convictions; just as novelists try to convey a sense of their story's verisimilitude through certain techniques imparting a sense of the concrete, the believable,[19] so science fictionists figure their data and ground their analysis in authority strings in order to connote the world's indubitable presence, hence lawfulness. Authority and methodology enable science to tell its story well.

Writers forget why they employ these and other techniques as science becomes a literary norm. Anticipating their professionalization later on, my graduate students proclaim that real sociology conveyed in piecemeal journal articles is quantitative and employs literature reviews. They acquire professional identities centering in their acquisition of technical skills and familiarity with a narrow subliterature; they express their emerging professional identities in their dissertation as well as in the team research they do with their faculty mentors, who frequently add them to their grants and thus enable them to publish while still in graduate school. Science is a normative literary practice, an ordinary accomplishment that forgets its reason for being, indeed that probably never knew it. The elaborate training students receive in methodology and statistics is merely genre training, giving them literary skills with which to embellish their published research. One does not learn the deepest secrets of methodology and statistics but only how to use them—how to write with them; one does not learn a literature in its nuanced variegation but only familiarity with it, its names and dates. Elegant software programs allow one to put one's literature on a computer file and then rearrange it appropriately to suit the citation needs of the project at hand.

Norms reproduce themselves, especially literary ones. We learn how to write by reading;[20] the most direct road to publication in the major journals is imitation of the substance and form of what has already appeared. This is the nature of discipline; without reference to extant literatures, one's work will be off the wall, ungrounded. Discipline is learned, and then self-imposed. Science is a writing style, not an apparatus affording indubitable truths. It is a mode and relations of intellectual production, narrating texts in the context of professional networks and disciplinary affiliation. Like literature, science is style; style is acquired, internalized, duplicated. More than anything else, this explains why the journals are so homogeneous, publishing similar work in similar styles—piecemeal, quantitative, figured.

The duplicativeness of science as a literary norm—we write what, and how, we read—is justified on grounds of disciplinary advancement, which is seen to proceed through accumulation. Believing this about what we do, in the meantime we keep busy, burying ourselves in subliteratures authorizing our narrow work and acquiring new methodological techniques as a logic of presentation. The usual left-wing complaint that journal editors screen out politically unsavory work is almost (although, it must be said, not entirely) beside the point; virtually no one does work that breaks the mold. Today the journals publish work by people using Marxist and feminist words, employing class and gender as their "independent variables."[21] This both enhances the impression of disciplinary pluralism and coopts dissidents; instead of criticizing the epistemological basis of political and economic privilege, they do mainstream quantitative research, only reinforcing disciplinary norms, hence power. They do so because (they convince themselves) it is useful to belong to the mainstream; it opens lines of communication. Really, it advances careers, buying inclusion.

My study of journal science reveals not only a certain literary style but also, through that, explains the remarkable homogeneity of disciplines. Where science is subordinate to canon as a literary requirement, each article only adds value to the whole. Single thrusts cannot disarm discipline; they only strengthen it by joining it on its own turf and in its own terms. The editors, in the interest of ecumenism—public relations—plead for articles using alternative theoretical and methodological perspectives. But they mean difference within sameness—authorizing literatures, different methodological techniques, different hypotheses, but not the abandonment of the whole positivist literary genre. Mainstream colleagues sometimes exhort me to send my work to mainstream journals. They neglect the obvious fact that editors of these journals will not publish it in its present form. I would have to rewrite it, thus changing it too much. They say they want me to "broaden" my work for a larger public. But the number of readers of mainstream journals is incredibly small—a few thousand at most, frequently only hundreds. To address a much larger public—the public—is methodologically illegitimate given the esoteric nature of science. Public discourse is nonscience, exuding poetic, even political, motive.

Many "radical" critics in the university complain that the leading journals do not have a place for them; they want to belong and thus to identify.[22] But there is a place for them if they canonize, figure, and number their arguments. The price of membership is low when even modestly competent people can iterate the dominant literary style of the journals. Even graduate students do it. The point is not that the journals are ideologically selective (which they are, too) but that the nature of ideological selection is not patently political but epistemological and ontological—the very theories of knowledge and of being. Positivist writing excludes the possibility of utopia; the search for laws disqualifies different history.[23] Positivism as a political language game encourages people to accept and enact their own oppressions by the "independent variables" of class, race, gender.

Positivism is the deepest ideology when it appears innocent, as in the pages of science. Mere representation as an epistemology presupposes a world that can be reflected at all. But a dialectical history eludes its figuring into invariant pictures and patterns. The text willfully embraces its corrigibility as a perspectived account, even an explosive text of political desire. That is to say, every writing is argument, even—especially—science.[24] It argues for a conception of the world, and thus a world, it would bring into being as a result of its publication. Austin is right; speech is action, writing above all.[25] Positivism is unique among rhetorics in stilling its own authorial, hence political, self-consciousness. It does so out of cupidity; it wants to support a world and does so quietly, figuring and authorizing it in the cold prose of representational knowledge.

This assumes we agree on the nature of science—to seek invariance and thus to predict with perfect confidence. That is a positivist version of science, not all possible cognitions. I am an empiricist in that I talk about the world; this book is an empirical study, nothing less. But it is also a theory of being, even more a recommendation for an order of being. I root democracy in dialogue, dialogue in literary openness.[26] These are slogans already. I am politically against science passed off as pure reflection; nothing only reflects without also constituting. The social world cannot be frozen into ice; every attempt to do so lowers the temperature, but never to absolute zero.[27] Thaws are possible. In rewriting science as a literary accomplishment, a script,[28] I weaken it politically when I show its subtle political agenda—to reproduce the given purportedly as unalterable. In this, science lies. While science is fiction, fiction can tell the truth, especially when it admits, even celebrates, its own corrigible nature—just another act of work or play.

A conservative, not a radical, first suggested that scientism is a secular ideology.[29] This is truer than ever, even on the left. It is conventional wisdom that argument must be supported by figural and statistical evidence, authority strings, elaborate notes and bibliographies, and dispassionate third-person prose written in the passive voice. The left in universities succumbs as readily as the right—sometimes even more readily because those on the left have been on the outside for so long and now want in. The alternative is not antiscience, however; Luddism is still wrong, trading bad history for miserable prehistory. I insist that dialectical literary strategies—sciences—are possible, even if I cannot demonstrate this by marshalling many Marxist or feminist examples of this in support. Objective analysis is not exhausted by the positivists; indeed, I would say that it is not attained by them.

Number, figure, authority strings are but conventions; they are more than that, too, having a subterranean political purpose in freezing the apparent present into the whole of history. As conventions, they can be changed, overthrown. They are choices; we can choose differently. Journal style is not required of science; it exhausts science by reproducing itself through the circuitries of power—control of journals—and imitation. Where many Marxists typically emphasize gatekeeping in explaining unequal access to media, I would place at least as much weight on imitation; scientists write science the way they read it, having no credible alternatives and clamoring to belong. I do not idealize; even if people essayed non-positivist science, they would not get published. But in the absence of qualitative alternatives to journal science, indeed with its proliferation even on the left, there is no alternative to the mainstream. Things flow along—careers are made, norms learned, power protected.

Chapter 3 The First Page

Science is written not to resemble a literary text; that much must be remembered throughout this book. Thus the challenge is to subject science to a literary reading, hearing it as if it were indeed the authorially driven narrative it is. Science conceals its literary artifice in order to seem to reflect nature without presuppositions or prepositions. The more the muddying effects of authorial perspective are removed, it is thought, the more convincing its account will be; this is called objectivity in the philosophy of science. But I say that this is a political project where the concealment of a text's literary nature dupes readers into living a conception of the world imbedded in seemingly authorless text and figure. Positivism is the name I give to science where it distances itself from poetry on grounds that it is the truer reflection.

Science's aim in its textless text is to suggest a conception of the world (here the social world) governed by invariant laws. Thus its veiled conception of the world would become a world itself through the readerly practices inevitably translating themselves into various sorts of renunciation. A text depicting a seamless, lawful world disqualifies a host of actions on the part of readers. While the relationship between text and world is indirect inasmuch as it has different effects on different readers, I want to understand the text's relationship to the world it describes as a deeply social—thus political—one.[1] When the sociology of knowledge labors to contextualize knowledge in terms of its antecedent social and economic influences, I want to address science's textuality as a political relationship among writer, readers, and the world that science would thoughtlessly reproduce by offering definitive accounts.[2]

Thus science's literary problem is to convince readers that its account is accurate, indeed that its pages are somehow divinely copied from a pregiven cosmos—not authorial accomplishments. Of course, people who write science know differently because they understand the sweat that goes into research and its literary scripting. They do first drafts, revise them, show their work to colleagues, revise again, and then send the work to editors who provoke further revisions. Science as a literary accomplishment is an endless series of negotiations between text and readers, the people Merton called gatekeepers.[3] Even assuming we have satisfied ourselves that our work is more or less acceptable, we almost inevitably find ourselves doing busy literary work before the pages see the light of day.

While we, as writers, know that science is arduously crafted and thus has a certain contingent quality—after all, the revisions are inevitably arbitrary and could have been done differently—positivism conceals this contingency as a rhetorical, thus political, posture.[4] Science is not supposed to read like persuasive argument, even if at root it seeks to do nothing more than persuade and thus build consensus. Science as an ontological posture fears unmasking as mere rhetoric, a particular arrangement of prose and figure on pages delivered as if they were merely copies of the world's indubitable presence. Although in my concluding chapter I argue for a concept and practice of science that shamelessly admits, even celebrates, its literary nature, in the intervening chapters I want to undo science's postured presuppositionlessness. Before I essay a different version of science, I will read it differently from the way it intends to be read—as objective fact having the figural authority of high science and the canon buttressing it.

Science Summary

In science's case, we both begin and end at the beginning. The first page of science articles in mainstream sociology journals combines prose, typeface, and the use of space to establish the article's authority and scientific worthiness. Far from merely a beginning, the first page of science announces its definitive address to a particular topic by established authors in the field. It does more than that. Most major science journals summarize the articles (frequently in italics) just above the body of the text and below the title and authors' names. In a certain sense, then, we can choose either to begin or end at the beginning. Impatient readers need only scan the italicized blurb (around 100-250 words) to consume the article's findings; or they can sample the findings summarized here in order to decide whether or not to plow through the main text.

The italicized "abstract" of the article invites beginnings and endings in its own terms.[5] It also serves subtly to buttress the scientific authority of the results and methodologies reported in the main text where we may believe that science can be summarized and thus communicated tersely. Although books lack these terse summaries, they are hallmarks of science inasmuch as they support the notion that science reports "new knowledge," findings, results, information, data. While it is uncontroversial that science does indeed report the world, the science summary with which most articles open (and sometimes close) reinforces the reporting nature of science. By being reduced from ten or twenty discursive pages to a hundred words of compact prose, science establishes the credibility of itself as objective description, indeed representation. The point of science is the data, the fragments of the world brought into the text as number, figure, prose. Article abstracts add value to the notion that science reports, reflects, represents.

This is not to say that article abstracts are particularly readable. For nonexperts in the particular subfields within which writers write about their topics, the dense prose of the abstracts can be forbidding, provoking readers to stop there. Yet it is not assumed that the different articles published in each number of science journals will be read and digested by everyone; the articles are addressed narrowly to experts. Others may read them or glance at their summaries, but more in the style of leafing through the thick Sunday newspaper. Science is canonical, working within established bodies of knowledge—literatures, discourses—familiar only to others who regularly plow that turf themselves. It is clear to any working scientist that few colleagues actually read the major journals as they read fiction, biography or even the Sunday paper. The esoteric nature of science articles, assumed from the start, is reinforced by the science summaries decorating the top of the articles' first pages. That they are unfathomable to many only underlines their verisimilitude.

> This paper asks whether...With these models we, test...We discuss some implications...for...[6]
>
> This paper critically evalutes...Drawing on...we find support for...Yet, contrary to the...[7]
>
> We use the...perspectives to examine how...The results of our analysis suggest...and thus should be integrated in future studies.[8]
>
> This paper formulates and tests...Using this theory...the article demonstrates...[9]

These examples of the summary abstracts appearing near the top of the first page in science articles indicate the descriptive nature of the accounts given. Terseness is valued, indeed required, in the short space allotted. In only a few cases do the authors use first-person privilege in discussing the authorial intent of the article summarized. Usually terms like "this paper" replace an active acknowledgment of authoriality.

As well, the tone of these summaries is conclusive. "The results of our analysis are" and "the article demonstrates" leave little room for doubt. Later we shall see how this conclusiveness is contradicted by the extraordinary hesitancy of the concluding "discussion" sections of most papers, where a methodological circumspection takes over. The science summaries summarize indubitable science. The results are reported as such; data are said to speak to the reader.

The phrase "The findings have implications for"[10] suggests an incontrovertibility to them. The results of the articles are the results, the self-evident implications of "findings" and data. This is a major stylistic strategy of science when it is crucial to disembody and deauthorize the prose presentation of data. Positivism assumes that the world can be made present in the text, speaking without the aid of words. Accordingly, the articles produce results, findings, information and not interpretations. Science appears to take no literary license in its representational posture; the findings are the findings and, as such, require no further interpretation. Of course, no one who writes science believes this. At the least they would admit that data bear a number of possible interpretations (even if they do not learn from this insight that science is literature—the conjuring of a world).

It could be argued that summaries published near the top of the first page of science articles serve the useful function of telegraphing the article's content for readers unsure about whether or not to read the article. It might be said that they save people time by truncating the results into a few pithy constructions. Summaries also suggest that science can be summarized, an assumption utterly uncontroversial for most working scientists. The problem with this assumption is that it regards the text of science essentially as a utility, not a performance valued in its own right. Expression gives way to communication when we make this sort of assumption about the nature of science. Knowledge comes in units, segments, communicable in summary form. Although trivially true when one can make representational claims about the world (e.g., it is 86 degrees outside today; the mean income in the United States is $26,000 per annum), this assumption degrades the performativeness of science as a narrative medium in its own right. Instead, again, we reduce science to its findings, the iterable claims together constituting a discipline's knowledge.

All that matters is text, not also subtext. Science's text is its findings when its subtext constructs a world out of those findings in a way intending to persuade. Both are important; but positivist versions of science, dominating mainstream journals in natural and social science, neglect subtext entirely in reducing themselves synoptically to terse, hard lines representing findings, data, results. But science is also the way it sits on the page, using words and figuring information into a believable whole. One might say that science reports and interprets but does not evaluate or recommend. Yet a nonpositivist version of science refuses to separate science's expressive, evaluative, and communicative natures, especially when it recognizes that the world conjured by science out of contingent scraps of data, figure, and number could as well be different.

Science is incomplete when it neglects its own literary methodology—the way it chooses to inhabit the page with gestures disclosing a world, indeed a whole theory of being. But science inevitably gives an account of itself when it accounts for the world, even if it tries to suppress that account in the interest of methodological rigor. A theory of knowledge eschewing its own theoretical character, positivism is the most theoretical position of all when it refuses to surface its own arbitrary assumptions about the representationality of knowledge. After all, abstracts, titles, citations, acknowledgments, graphs, tables, and figure are only conventions; they are inessential and could be different. In failing to make explicit its own choices for these accouterments of presentation, science conceals the corrigibility, indeed political purpose, of its account. As such, then, this mode of science lacks objectivity when it fails to include itself in its own account of the world outside of it.

It is not that article summaries are impossible—only that they are poetic transformations of the whole text, as such changing it. People who write science impatiently pen their abstracts once the whole paper is written, frequently failing to consider the weight they add to the text itself. Of course, there is no text itself; the summary belongs to it too, and frequently changes it, especially when it suggests that the text underneath it be read as dumb representation, not also as passionate theoretical argument or simply fiction—writing that constructs a world. Writing the summary assumes that the article can be summarized, indeed that it should be. It helps the article lose whatever self-consciousness it might have had to begin with. By suggesting a certain representational reading, the article summaries ensure that reading, a definite literary choice when other choices are possible.

If article summaries are read only to indicate the authors' sense of what is important about their argument (admitting that they are, after all, engaged in argument), that is one thing. Indeed, it opens up the narrativeness of the text in inviting dialogue—a world that could be written, and thus lived, differently. Yet the summaries are not embellishments of self-consciousness, retrospectives functioning as introductions, but simply reports of reports. Whereas the body of science's text lacks much self-consciousness about its literary (and thus political) motive, the summary at the beginning is the antithesis of self-consciousness. Science gains legitimacy by presenting itself as a report, a representation. The summaries oppose literary versions making thematic their own contingent, even self-contradictory, quality. As such, they are enemies of dialogue, suggesting that the reports they summarize are unavailable to refutation, only inspection. Self-contradictorily, this violates positivism's norm of replicability, the assurance that methodologies can be repeated by other observers and thus checked for their validity. But almost no one replicates in science; it is too time-consuming and fails to make careers. Thus summaries ironically function to make science into a monologue, a selfsame account of a frozen world.

While it is possible to summarize, in suggesting that science's summaries serve subtly to close off dialogue and thus deter challenges to it I am not recommending other kinds of summaries, except perhaps as moments of textual self-consciousness. Authorial self-interrogation could happen anywhere in the text, and it should. Article abstracts are merely conventions of science's logic of presentation today. Indeed, summaries beginning articles are extremely strange on the face of it. They would seem to belong at the end, after the narrative work has been done. Instead, the end of science's articles are called "discussion," usually the last page or so where the authors discuss the implications of their reported results. One might otherwise expect the discussion to come at the beginning or in the middle instead of the end, just as summaries should summarize what has gone before.

These are literary norms that protect the science text against counterarguments. Whereas fiction presents itself unproblematically as a true account of an essential world, science, as a peculiar kind of fiction, differs only in that it must claim validity methodologically, showing its credentials to make a certain case.[11] Summaries at the beginning and discussions at the end are such conventions, although ones designed to still debate by enhancing claimed authority. But readers and writers of science forget that science's methodical self-presentation is nothing but a norm; instead, they mistake these stylistic conventions for the real thing—Essential Science, science as-it-must-be. Thus one article in imitating another only adds value to the norm, virtually ensuring that it acquires a binding force independent of its original reason for being—to establish validity for discourse.

Science refuses the notion that it *is* discourse; it sees itself merely as representation, the report of findings through prose and figure. Summaries appearing at the beginning of articles reinforce the article's scientific authority and close off dialogue; as reporting, science dispassionately tells it like it is. The passive voice of the article summaries lends support to this—"An evaluation of three models...is made..." Nothing literary seems to be going on. Whereas the text is dead prose, merely reflecting a given world outside it, the summaries are even deader; by seeming to have no author, they embody authority. The summaries are usually the first piece of prose that readers read. As well, when they begin the article for many readers, they also end it for many. The article becomes the findings reported in the blurb atop the opening of the article's text. There is no article, no other text, but only the report of findings. In this sense, the summaries perched atop the first page effectively negate the text beneath it even as they appear faithfully to report it. With the report at hand, one need read no further.

Science summaries conceal their own contribution. Although one can truncate any text into less than it is, doing so makes it more than it was; summaries chill science into mere description when, in fact, descriptions essay the world—what is, what is possible, what should be.

Topic

When the abstracts are often the first (and, ironically, the last) prose really read on page 1, the very first thing that the eye notes is the title, often printed in large, boldface type, as such set off clearly from the other marks on the page. Titles carry baggage not only in what they say about the piece beneath them. As important as the "content" they suggest in what is to follow, they have inherent significance in that they stand prominently above all the other prose on the first page—big, bold letters literally having the first say about the words to follow. And they are written in dryly descriptive language: "Suburban Socioeconomic Status Change: A Comparison of Models, 1950-1980," "Industrial Restructuring, Gender Segregation, and Sex Differences in Earnings," "Cohorts, Regimes, and the Legitimation of Democracy: West Germany Since 1945."

As such, science titles reinforce the science nature of the articles following them. They are not essays, theoretical musings, poetry; they promise a report, a representation, and they deliver (even as early as in the blurbs summarizing the articles directly beneath the title). Only rarely are the slightest literary embellishments allowed: "The Ties That Bind: Organizational and Class Bases of Stability in a Corporate Interlock Network." The phrase "the ties that bind" suggests a certain authorial skill in dialectic; but the impression is quickly corrected by a narrowly scientific subtitle. It is clear that titles serve to give title to the reports, descriptions, accounts proceeding under their name. The bulkier and clumsier they are, the more they convey the impression of no-nonsense science; they are anything but ethereal.

Of course, this is the stuff of jokes about stuffy, disciplined academics. Surely people who write so inelegantly are not good for much. Even scientists poke fun at the lists of obscure, tortured topics regularly appearing in their professional journals. Recent sociology journals include the following bizarre titles (and accounts): "Initiation of Coitus in Early Adolescence," "Further Evidence of Sibship Size and Educational Stratification," and (my favorite) "Changes in the Core of the World-System and the Production of Utopian Literature in Great Britain and the United States, 1883-1975." They could as easily be about "Zucchini, Artichokes and the National Debt: A Cohort Analysis, 1973-1976." But these titles suggest not only content; in their obdurate density and narrowness they promise contributions to existing bodies of knowledges—literatures, as scientists call them. Their weird narrowness is excused precisely on grounds of their legitimacy as disciplinary advances. More sweeping, and thus more readable and interesting, titles would undermine the careful accumulation of expertise about the age of first coitus, sibship size, and utopian literature. As scientists—which their titles say they are, must be—the authors enhance their professional credibility by sticking to narrow topics conveyed in the science prose of sheer description.

The titles, in effect, say: This subject is important to a certain subliterature in the discipline, no matter how boring or trivial it appears. The issue is not simply literary stylization; hot titles can conceal cool, dull prose underneath them. By the same token, some of the world's greatest writings have had ambiguous, unpromising titles—*Republic, Capital, On Liberty*. Indeed, it is a stock in trade of "fast capitalism" (see my *Fast Capitalism: A Critical Theory of Significance)* that bad work is dressed up in flashy clothing in order to move quickly in the emporia. One is rightly suspicious when one comes across intellectual work that seems too facile, trendy, glitzy. Yet work that revels in its own obscurity and technical nature is not better in its smug professionalism. It merely establishes authorial credentials in reproducing the authority to which it would make its tiny contribution.

Science to be science, then, must appear both dry and narrow. Science titles suggest a certain studied ponderousness. Everyone but the expert is bound to shy away, perhaps lingering only long enough to peruse the abstracts printed below the title. This ponderousness is a methodological requirement of science where it has become a norm. This is not to say that slicker, jazzier titles would solve the problem of the articles beneath them; after all, obscurantist titles share obscurantism with the texts that they title. Articles in science are totalities of nonsense. One can read titles alone and know what is happening in the articles beneath them. As minitexts in their own right, they help establish the article's scientific credibility, especially when they are the first things readers read—published on the cover and title page and on the first page of the articles themselves. I am interested in the first page of science because it is there that the reader finds the telltale traces of a version of science silencing all competitors; titles merely establish part of the article's credibility as legitimate disciplinary work. Like science abstracts, they contribute to the overall sense that the article about to begin is thoroughly disciplined—narrow, technical, sober, constructive, descriptive, a good report. Showier titles would raise questions about the article's validity as sound discourse.

It is also important to note the relationship between the title and the authors' names beneath it. Whereas the science blurb is usually printed in a different typeface (frequently italics) from the body of the text, thus suggesting their differential statuses—blurb higher than main text—so the title usually dwarfs the tiny names of the authors and their institutional affiliations printed beneath them. This is odd only because it could be different. The title of Bob Woodward's recent book about the Central Intelligence Agency (CIA), *Veil,* is dwarfed on the dust jacket by Woodward's name. The book is Bob Woodward, who also happens to be the author. The publisher shrewdly recognizes that Woodward's fame will sell books; that he writes about the CIA is only incidental. Although this is an obvious marketing ploy, it also suggests the book's literariness, its crafting by a real human being known to the public as an *auteur*.

Science diminishes the author (now, typically, the authors—a whole team of assembly workers) by placing her or him or them in tiny print underneath the bold, large title. What matters in science is the technical topic and the authority authorizing it, and not the particular writer or writers working on it. Indeed, science's legitimacy is enhanced the more authorial presence is obscured by the meshing apparatus of prose, figure, and number, together squeezing out the writing's literariness as an ontological—hence political—strategy. Articles could be written by anyone competent in the technical method appropriate to the topic and literature at hand; the article on age of first coitus stands on its own as disciplinarily legitimate work accomplished by competent but nearly invisible professional experts.

Science ironically deauthorizes its account in order to give it authority. The less present the author as an individual known for his or her prior work, intellectual and political style, temperament—Alvin Gouldner, Allan Bloom, Bob Woodward—the more reliable the account will seem to those who value positivist presentation as a logic of representation. The more present the author, the more suspect as good science the piece will be; it will dangerously veer toward rhetoric, polemic, even fiction—the worst charge of all. The title overwhelms the author with its obscure density precisely to reduce the role of the author in having written the piece of work to which it lays title. Where the savvy salespeople in commercial publishing understand that Woodward is bigger than the topic about which he writes, so academic editors and authors recognize that the efflorescence of authorial subjectivity should be kept to a bare minimum. At stake in both versions is textual credibility.

The title in science is extraordinarily wordy in order to disclose itself to other competent experts who work its field. Yet this wordiness is not an embellishment, a play of concatenating signifiers trading off one anothers' senses, but an encyclopedic string of adjectives and nouns that function merely discursively. The titles of my last three books (including this one) have averaged 1.7 words. Science titles are long and dense. "Urbanization and Underdevelopment: A Global Study of Modernization, Urban Bias, and Economic Dependency," "Age Structure and Crime: Symmetry Versus Asymmetry and the Projection of Crime Rates Through the 1990s," "Understanding Labeling Effects in the Area of Mental Disorders: An Assessment of the Effects of Expectations of Rejection," "Changes in the Core of the World-System and the Production of Utopian Literature in Great Britain and the United States, 1883–1975." These titles never seem to quit. Packed with words, they lie lifeless on the page. They eschew encyclopedism (the bane of disciplined science) but nonetheless proliferate volumes.

In this sheer density of words, the titles further bury the literary caprice of the particular author. Wordiness specifies and narrows. Under this heavy weight, one can but write about the narrow particulars of crime rates and mental disorders. Whereas in literature anything goes, in science virtually nothing can happen outside the delimitation of the cumbersome title. Indeed, many titles contain subtitles further narrowing the scope of the effort. When, occasionally, the authors begin playfully or popularly—"So Happy Together?"—they quickly narrow the topic in the subtitle—"The Impact of Gender Segregation on Men at Work" (and this is an example of a title promising a relatively broad treatment). One might observe that the lengthy titles are really parsimonious in that they communicate effectively, promising a specific treatment of a particular topic. But this sort of specificity burdens the author with the expectation that he or she stick to the local issue at hand and not overstep the boundaries of the title, speculating "beyond" the data. Indeed, this issue looms large in chapter 8 where I discuss the tenative nature of science conclusions.

It is common knowledge that advanced science is narrow, exploring the depths of topics frequently treated. There is nothing inherently wrong with this. But I am concerned with the rhetoric of narrowness as it becomes a norm followed unthinkingly. Although some subjects demand and deserve fine-grained inspection, science's titling practices effectively disqualify more global approaches in all instances. Nothing gets published that does not present itself in the leaden, byzantine titles announcing the reporting of information. Mainstream science does not title itself "Truth and Beauty" or even "fast capitalism" (one of my book titles). Fine-grained approaches dominate simply because they have dominated; in turn, they require titles and articles to reproduce this historic narrowness in the interest of disciplinary legitimacy. The truism that too much science is narrow is nevertheless true; but narrowness is a literary practice followed slavishly and not an essential requirement of science as such. Artfulness intending writing full of both author and topic—indeed, making their connection thematic—is disqualified simply as nonscience—speculative, ungrounded, suitable perhaps for literary journals (although they, too, are increasingly dominated by a science model of presentation).

This is not at all to deny that big names have privileged access to publication in journals and books. The typical Marxist focus on the gatekeeping functions of editorial work is correct as far as it goes; I merely give equal attention here to the literary practice of hegemony.[12] I want to understand hegemony as a normative and narrative practice relying heavily on mundane literary techniques having to do with the arrangement of type and prose on the first page among other things. While authorial subjectivity in general is virtually submerged in science's representational text, buried by title, abstract, literature review, methodology, and the figural display of findings, particular authorial subjectivities – flesh-and-blood individuals – still hold sway in particular disciplines. The *ASR* even published an extraordinary article recently proving, in the hesitant language of statistical inference, that big names from prestigious departments have unequal access to publication in that journal. We on the left knew that all along; our database was ourselves: We could not get published. It is not trivial that certain people are published and others denied print. But that is not the whole story. Even big names have their names published in agate type under the ponderous headline of their titles. They are no less exempt from the methodological practice of self-suppression; they differ only in that they command more space in the leading journals, the crucial real estate of disciplinary advancement.

Titles constrain the paper's scope. They subdue the tendency of liberated writers to wander from one thing to another even if they are following the secret logic of the complex world in unraveling it. In turn, titles are themselves constrained by the extraordinary development of what I have called the various subliteratures constituting each discipline. These subliteratures, hierarchical networks of books and articles on particular topics within a discipline, constrain topics and hence titles by their very weight. The evolution of these literatures has momentum; at a certain time people will be working feverishly on some topics and not others. There will be established experts who must be acknowledged. Authors attempting to add their voice to research on a topic must work within the logic of presentation and extant body of expertise surrounding the topic.

Thus titles largely reflect the incredible specialization of research work in science. People write about topics in the context of established work on those topics. Thus titles in their narrowness seem to follow from the work already published on aspects of the topic at hand. Thus an article on age of first coitus will necessarily build on past research, which in turn will limit the article to certain strategies of presentation, method, analysis. Specialization not only militates against broad-gauged inquiry; it also implies an obedience to canon that further constrains the writing people do. This issue looms larger in the following chapter where I discuss the way that established authority makes its presence felt in the text of science. Its first way of doing so is in the beginning; titles in their particularism connect with other literary approaches to the topic, thus narrowing further (and dictating a common style of intellectual investigation).

My final observation about titling as a normative practice in science is that it both reflects and reproduces a titled, disciplined world. As titles reflect the state of development and style of work in particular subfields of disciplines, so they add value to those traditions. Titles seem to come out of the blue in settling on the page in heavy ink. Insiders know that titles have history behind them; others have written within the same genre, and they will be recited in the paper's literature review as well as in the bibliography. In their seeming arbitrariness, titles establish their own importance as well as the enduring importance of that type of work. By announcing an article on "Suburban Socioeconomic Status Change" or "Using Adjusted Crosstabulations to Interpret Log-Linear Relationships," the authors authorize extant traditions as well as secure a place for themselves within that tradition. It is not intuitive that suburban status change or the use of adjusted crosstabulations are interesting, even legitimate, topics for sociologists. Indeed, I have been in the discipline for nearly twenty years, and I do not know exactly what these words mean—suburban status change, adjusted crosstabulations. Although I have heard the words used before, perhaps in different arrangements, they cry out for explication. I will have to read the whole article before I can begin to understand the title.

Obscure titles, in being published, perpetuate the obscurity. Not only does obscurity in general become a norm but in addition the particular title itself gathers momentum as a legitimate literary topic. One wanders through the tables of contents of academic journals searching for some method to the madness. One comes up empty; the articles announcing reports on densely specialized topics are not only unconnected but frequently indecipherable in their own terms. The disciplinary generalist fails even to understand the words in the titles, let alone the prose of the articles itself. I am not a proponent of ordinary-language philosophy that resolves all philosophical problems by an appeal to common usage. Yet one assumes or at least hopes for a sufficiently public language within which specialists can talk to one another about their work, not only across disciplines but within them. As it stands today in science, one cannot assume that one's colleagues across the hall will understand one's own work—even when they seem to share a discipline.

What is happening, then, in the production and reproduction of obscurity? Scholars are staking claim to space in the leading journals that they hope to translate into career capital. One of the problems with poststructuralist approaches to literary theory is that, with Derrida, they assume that the text has no boundaries, nothing beyond it. This is wrong. Science pages matter precisely because they are enmeshed in certain social relations of power, notably intellectual ones. Science's mode of presentation reinforces intellectual, hence ideological, hegemony; some people get more space than others. Where there are many journals (and nonacademic media of presentation), there are by definition only a few leading ones, representing and thus reconstituting the dominant mainstream. Ideology entrenches power as a norm. Thus literary practices both support and reflect disciplinary power.

Although it is necessary to notice the role obscurity plays in the literary reproduction of disciplinary power, some obscurities are more highly valued than others. Mainstream positivist sociology prefers the obscurity of quantitative method to that of European phenomenology, Marxism, feminism.[13] These other traditions are dismissed as nonsociology and thus denied places in the leading journals. This book examines further the other dimensions of mainstream obscurity in the chapters to follow. At this point, it is important to understand the role titling plays in the intellectual reproduction process. Lengthy, wordy titles authorize the analyses done in their name; they reflect what is current in the relevant subliterature; and they overwhelm the presence of the living author, denying the possibility of political perspectivity, passion, poetry.

Collective Authority in an Age of Teamwork

The third thing to notice about the first page is how frequently the author is more than one author. Following the practice of the natural sciences, social science articles are increasingly credited to coauthors, even whole teams of authors. In part, this reflects the disciplinary trend toward sponsored team research; senior faculty members who get grants add their junior colleagues and graduate students to them in order to share the work and assist them in publishing. Tenure and promotion committees currently have difficulty in weighing these authorial contributions. Do multiple-authored pieces "count" for as much as solo articles? If not, then how much currency should they bear? In part, the trend toward collective authorship also reflects the growing degree of the division of intellectual labor; one coauthor might be responsible for the literature review, another for doing the computer runs and preparing the analysis of data, and the third for compiling supporting references, entering the paper on a word processor and handling the editorial correspondence.

I am interested in the way that collective authorial naming constitutes a greater authority for the writing than could be claimed were the paper the brainchild of a single person. Authority diffused is authority strengthened. It is implied that a team of writers can do better science than a single scientist, if only because the team check one another and catch mistakes. While tenure committees may have trouble evaluating people's respective contributions on team articles, the fact of team authorship itself adds credibility to the analysis, especially when the multiple authors are well known and have posts in prestigious departments. Shared credit may be the price paid for the enhancement of authorial authority in an age dominated by the science text. Indeed, the most enlightened tenure committees, recognizing the shift in social science toward a natural science model of collaborative and quantitative research, accord team articles the same weight as solo efforts, perhaps in a subtle sense even more weight.

While prominent titling overwhelms authorial subjectivity, that subjectivity ironically supports its own disempowering in adding authority to the article by suggesting a thoroughly contemporary mode of scholarship. Team articles are better than solo articles in that they are more likely to be quantitative, funded externally, and valid; the division of intellectual labor they represent is often associated with a discipline's increasing specialization, generally regarded by positivists as good in itself. Although there is nothing inherently wrong with working together (after all, intellectual democracy might spell the real thing), multiple naming is highly characteristic of a science ruthlessly expunging traces of its authoriality at every turn. This is a major irony for science; the more names listed on page 1, the less subjective the results appear. Or perhaps this only appears to be ironic. After all, the listing of multiple names suggests cutting-edge science grounded in the mutual correction processes undergone in shared authorship.

Whereas multiple naming suggests an article's authority by removing the potential impact of a single temperamental author, authors who write from different institutions strengthen authority even more. Incestuousness is reduced, and objectivity thus enhanced, when many eyes from different schools contribute to the same analysis. Of course, this may be deceptive where people may have once worked together as colleagues or in teacher-student relationships. But appearance is all important here, as everywhere in science's text. Multiple authors from different prestigious institutions have more authority than multiple authors from less prestigious institutions, from the same institution, or than solo authors—the sorriest souls of all.

The real worry is that solo authors might go off half-cocked, proposing theories and doing analyses scandalizing the established literature. The more intellectual work is done collectively, the less writers will freelance beyond the normative boundaries of the discipline, even beyond liberal reasonableness in general. Of course this is true, as anyone knows who has ever collaborated in their intellectual work (especially for those who do mainstream work or at least want to see it published in mainstream organs). One's coauthors domesticate the argument, ensuring that certain standards will be met. And we derive these standards from precedent—what the journals and publishers publish already. If virtually every paper in the major journals proceeds along the lines that I reconstruct in this book, then it is unreasonable to deviate, especially if one is not a star and thus lacks certain advantages at the start. One's coauthors usefully caution one about large mistakes of method and form, as well as about smaller issues involving the substantive literatures within which one is working.

Editors know this. Thus they favor coauthored papers as evidence of intellectual disciplining. One can count on a trio of authors from Wisconsin, Michigan, and Indiana (major departments of sociology in which well-known scholars work in the tradition of midwestern empiricism); they are likely to have paid close attention to issues of disciplinary form and content. They are sure not to be high-wire acts working without a net. Instead, they will have considered carefully the impact of their work on prospective manuscript reviewers, thus saving much time and effort later on. And indirectly, by conforming to what they take to be methodological and substantive correctness, they promote that conformity in others when they succeed in having their work published. This is not to say that every trio from major empiricist departments will have their work accepted (after all, acceptance rates are frequently at or below 10 percent of all submissions),[14] but they certainly have an edge over the lonely author who submits from the University of Southern Nowhere.

I have noted how authors' printed names on the first page of journal articles are almost inevitably followed by the listing of their home institutions (with no mention given to department). The journals assume that everyone who publishes is an employed sociologist; unemployed philosophers are not invited to submit. Although this assumption is frequently wrong on its face when the sociology job market is tight or when interdisciplinary research may be housed in departments outside of sociology, in itself the listing of institutional affiliation reproduces itself. That is, we assume that people who publish in mainstream academic journals should be employed academics narrowly located in a single home discipline. When this is frequently not the case, we tend to devalue their work. The linear literalist might respond that the listing of institutional affiliation only performs a service, informing readers about where a scholar works and thus implicitly encouraging reader-author contact in the future. But that is given the lie by virtue of the fact that journals rarely print the name of the academic department to which people belong. Instead, people are assumed to be from a single home discipline—here, sociology.

Although it is true that certain information is imparted when authors' home institutions are printed on the first page of their journal articles, the real purpose of this practice is to divide outsiders from insiders. Insiders are those people employed as sociologists at Florida State, Chicago, Texas A & M; outsiders are anyone else, particularly people outside academia altogether. When, occasionally, people from think tanks and consulting firms are listed by their employment title, these are exceptions; and it is generally assumed that these people hold degrees in sociology. Titling and naming practices reflect and reproduce the discipline in its own image. Sociology is what is published in sociology journals by sociologists (who work with certain methods, etc.).

Acknowledgments

When people publish alone, they have to employ a variety of compensatory strategies to give the impression of intellectual collaboration and thus domestication. In particular, they need to lard the bottom of their first page with acknowledgments to famous people belonging to important networks who, it can be inferred, have given their blessings to the article in question. Most journal articles these days open with asterisked sections at the bottom of the first page expressing lavish debts to mentors, colleagues, graduate students, even the anonymous reviewers who appraised the manuscript for the journal in question. Although one might simply acknowledge these acknowledgments as gratuities, they legitimate the article in question, especially when it is sole authored. The more powerful people one can claim in one's stable, the greater one's authority. This is essential not only in securing journal space but also in rewarding one's mentors and colleagues for, in effect, agreeing to have their sponsorship acknowledged publicly.

The elaborate acknowledgment of others at the bottom of the first page of journal articles is a trend rapidly gathering momentum. Indeed, the same thing is happening with books. The acknowledgment section becomes a place for paying off the large numbers of people one has accumulated by the time one is a professional academic. As a literary device, these acknowledgments help prop the text at the bottom of the page. Whereas the paper would otherwise be weighted down with title in boldface, authors' names and affiliations, and then the abstract, acknowledgments give a bottom to the page. The first page of the text is merely interposed, almost irrelevant to the figural display of the page in its totality. It is clear that the acknowledgment section is growing in length, squeezing the text even more.

At the most obvious level, these acknowledgments indicate that people must increasingly acknowledge their debts in order to get published. The lists of name are not arbitrary; after all, one must stop somewhere so that the top matter and bottom matter do not utterly overwhelm the main text. Page 1 cannot be only embellishments; without substance against which to serve as counterpoint, embellishments by definition cannot embellish. They only decorate when they are diversionary, suggesting another text alongside the text that appears to dominate the reader's attention—the text that "says" the article, presenting the findings in the straightforward voice of science. The other text or subtext can be found in the margins and in the figures and designs increasingly littering the science page. Acknowledgments are a crucial part of the other text or subtext, especially as the space they claim on the page increases with the growing importance of the citation of one's networks.

Acknowledgments in science suggest validity, authority, and domesticity. Wild writers do not see the world clearly; they rebel and thus cannot be trusted; and they do not have friends. The more names (especially "big" ones) one can cite at the bottom of the page, the more trustworthy one's account will seem to be. This is especially necessary when one publishes alone, without the buttress of coauthors from diverse institutions. But acknowledgments do not just acknowledge; they also put prose to the names listed: "The paper has benefited from comments by"; "We thank"; "We are grateful to"; "for their helpful comments"; "whose invaluable contributions are gratefully acknowledged"; "for computer analysis"; "for his helpful advice"; "for their methodological suggestions"; "The author would like to thank...the editor of *ASR*."

Let me pause here. The last acknowledgment is especially telling. A single author must not only thank scholars in the field but also the editor of the *ASR*. Now surely the editor has a hand in all decisions to publish, as well as in the editorial work supporting revisions made to the paper. Virtually nothing is published as such; so-called revise-and-resubmit decisions nearly always precede eventual publication. But in this instance the author felt compelled to thank the editor, especially when he was working alone and could not sport a large cast of supporting characters. The other remarks more or less fit the same mold. They thank people who have either read the paper or had something to do with its gestation for "their helpful advice." The word "helpful" is key here. The advice was friendly, nonadversarial. It was tendered in the spirit of improving the article, and it was offered by friends. That one has friends to whom one can turn for this sort of advice is a crucial part of the text of acknowledgments. It means that important people spent the time to read one's work or at least talked to one about it. Subtly it conveys a certain authority.

Many papers also thank the anonymous reviewers used by the journal's editor in order to appraise the initial and subsequent drafts of the article. This is bizarre, especially when the author cannot know the names of the reviewers in order to list them alongside the other people named in the acknowledgment section. Why thank people one cannot name? One might dismiss this as a folkway, simply repeating what others do, but that is not the whole of it. Everything means something, even if it is something else. Science is peculiar in that its figural displays and marginal embellishments frequently count for more than the text. I read outside the margin of the text where I search for significance; the disembodied, stultified prose of science makes it difficult to find much life and thus significance outside the boundaries of the prose page itself.

Anonymous reviewers are thanked for two reasons. It means that one is professionally graceful, giving thanks where thanks are due. In a dog-eat-dog academic marketplace the author still has the civility to pay debts. It also means that one is doing science; anonymous referees are the backbone of academic publishing. Key arbiters in the marketplace of ideas, they allow objective knowledge to be appraised and then disseminated. Without fearing the disclosure of their names, they can evaluate manuscripts rigorously and thus ensure that journals and publishers publish only high-quality work. The idea behind this publishing norm is that outside referees trained to have a certain expertise in one's field will be able to replicate the reported results themselves, or at least to appraise one's stated methodology. If two or three referees agree that a paper needs further work, it probably indicates that the author has not spent enough time on it or simply lacks the background to give a good account of himself or herself.

At one level, it seems strange to name nonnames in acknowledgments. At another level, it seems redundant, especially when we all know that science articles are refereed before they are published. But at a deeper level, the acknowledgment of anonymous article reviewers reminds readers that they are about to read science vouchsafed in the usual professional norms and not the essayistic meanderings of a crackpot. While editors have absolute authority in assigning papers to particular readers and then in evaluating the evaluations once they come in, it seems important that people acknowledge that their work has been processed through the alleged marketplace of ideas, thus giving the work more legitimacy.

The acknowledgment of referees also implies that one has actually responded to the criticisms of reviewers in making revisions. Inasmuch as almost nothing is accepted for publication as such, one is inevitably expected to make revisions before publication. Thus one demonstrates one's professional responsibility by seeming to pay attention to readers' comments. A small gesture, perhaps, but one that does not go unnoticed. Sometimes authors even admit that their paper has gone through "earlier drafts," again underlining their obedience to the norms of collective science. Although one can do revisions without acceding to each and every one of the readers' suggestions, at least by acknowledging them at the bottom of the first page one appears to have done so. When we read science as a literary accomplishment, appearance counts for more than reality; indeed, there is no such thing as reality—only the strokes and marks reposing tentatively on the page.

An Observation about Science's Other Text

I began to observe above that science constitutes two texts: what the article says and what appears in the margins, beneath, above, and beyond the prose itself. Both are important; thus I read notes, bibliography, acknowledgments, titles, namings, abstracts, and design as much as the words of prose themselves. This is especially important when science purges itself of authorial presence. Articles written descriptively in passive voice reveal little of what drives them to be the texts they want to become; we must search elsewhere, in authorial traces, for the desire animating every literary text. This is why many readers turn first to a paper or book's acknowledgments (and then perhaps to notes and bibliography) before delving into the work itself. In science, sometimes the most telling things happen between—or beyond—the lines.

My study of average science convinces me that while text and subtext coexist, they exist increasingly in counterpoint. There is a real struggle going on between the main text's prose and the hieroglyphics of title, names, abstract, acknowledgments, figure, chart, graph, and equation—subtext. This struggle is over space itself: who gets what in the way of disposable paper in the journal article. While the main text needs a subtext in order to buttress its scientificity, by differentiating subtext from text science sets up a struggle between the two that threatens to overwhelm prose with all the marginalia it needs to make its case convincingly. When the subtext gives authority to the text, it has a logic of its own, relentlessly expanding to fill the available space. That is why bibliography, acknowledgments, and figure have expanded so much in recent years. One would almost conclude: the less prose, the better.

Both text and subtext can be read for significance. They must be read together when they constitute a single whole. But it must not be assumed that the subtext is simply functional for the text, only giving it a science aura desperately required for disciplinary legitimacy. The subtext also competes with the text when its gestures become fetishes. My argument is that they do especially when these gestures—abstracts, titles, acknowledgments, figure, math—become norms without which writing is deemed illegitimate. Soon the gestures designed to validate the text become text themselves, squeezing out the prose that was supposed to have been the point all along. The accouterments of science are reified into science's essence and thus all articles to be published must stuff themselves full of the mumbo jumbo and subtle codes reposing outside the prose itself.

While abstract, title, and acknowledgments are indeed narrated in prose, they have a different significance from the main prose. This differentiation is revealed to the most cursory glance: Subtext almost always takes place in different typeface from the main text. To signify, it must be set apart from the text's "body," as it is called. In the case of the abstract, frequently italicized, the prose is compact and dry, utterly purged of authoriality—even more so than the main text. Although not chatty, acknowledgments give more hint of the authorial subject; she or he has colleagues, mentors, friends, as well as a mailing address.

Page 1 of science articles presents a blur of styles, typefaces, and sizes. The main text seems almost incidental to the busy activity flying around it. It is squeezed between top matter and bottom matter, gaining sustenance from them. In fact, as we see shortly, the main text of page 1 is itself usually littered with parenthetical names and dates in order to create even more legitimacy for the article. The busy technical activity comprising page 1 is an antidote to science's essay character; the more typefaces, type sizes, names, number, dates, parentheses, the less science appears to be simply a flight of fancy—even though that is what it always is. The top matter, bottom matter, and figure coexisting with regular prose on the first page provide it with an infrastructure, a foundation. Although the use of heterogeneous forms and styles on the opening page seems to suggest not only differentiation but also hierarchy between main text and subtext, in fact the hierarchy is curiously reversed; the text could not exist without the other embellishments of what is taken normatively to be science. Although the real point of science articles appears to be the text, it creates this impression by feeding off the top matter, bottom matter, and figure holding it in place.

Science, to be science, needs to negate nonscience. It needs to exist in counterpoint to an outside to which it is inside. Thus the essay character of acknowledgments and the unsupported declarativeness of the abstract heighten the validity of its main prose simply by being different from it. The marginal and figural embellishments of science act poetically to differentiate themselves from the nonpoetic substance of the work in question. At the same time, dialectically, these nearly poetic signifiers existing outside the text proper support its claim to be scientific. Circularly, science is what science does. Where abstract, title, acknowledgments become a norm of science, then it is impossible to dispense with them when writing science. Indeed, in being distinct from main text in their nearly essayistic nature, they suggest that the main text is superior in its studied rigor.

Although I use terms like *text* and *subtext,* I do not really think that there are two different texts in science. There is only one: the complex totality comprising all the rhetorical features of the first page plus the elaborate use of figure and prose in the balance of the article. Deconstructive work reduces text into its interconnected components only to make a surprising point: Differential reading is impossible. We always read wholly, failing to segment the page into seemingly extraneous embellishments and vital prose constructions. Precisely because we read this way, we must slow down our readings in order to understand how writing destines a certain reception. In skipping over the embellishments, we do not miss them entirely; what they say about themselves and then indirectly about the main text to which they exist in counterpoint is never lost. The point sinks in. This is science, supported by the normative rhetorics figuring the page both inside and outside the margins of the text.

Science deceives us that there are, in fact, important and marginal aspects to its mode of presentation. It would support the notion that acknowledgments, abstract, title, affiliation are all subtexts, not real representations. But everything on the page is a text designedly figured to create a whole version, especially one defying its literary reading. Science can claim representationality precisely because it uses marginalia to support it. Its coauthored, institutionally affiliated, networked, abstracted, canonized argument suppresses its poetic nature—the fact that it has been written by corrigible authors simply conjuring a world out of nothing. Positivists insist that their work adequately reflects reality, but the truth is different. Writing only bears a relationship to reality; indeed, as part of reality it helps constitute it, thus ensuring the world of which it only pretends to be a simple concept. That is the ideological nature of science. I want to inquire further how it reproduces that nature through the narrative practices comprising it.

Chapter 4 Authorizing Authority: The Literary Irony of Science

The Literary Irony of Science

In chapter 3 I talked about the way heterogeneous rhetorics constitute a convincing first page of science articles. Titling, naming, acknowledging, and abstracting all buttress the authority of the main text as an authentic representation. As marginalia, they uplift the main text into a significant medium of disciplinary communication; as (what people take to be) essential parts of science, they guarantee the work's scientificity. But something also happens on the first page inside the margins. The prose begins. In this chapter I want to address the typical first sections of science articles; they possess a striking homogeneity, indeed characterizing the beginning sections of papers, books, dissertations, and theses throughout most science disciplines. In these telling first sections the authors engage in a "literature review," as it is known, grounding the paper's topic in a distinctive disciplinary subfield and staking a claim for its specificity.

These literature reviews constitute a literary irony of science in that authorities are cited in defense of a tried-and-true approach to similar topics (e.g., racial prejudice, marriage rates, income inequality), and at the same time the authors claim a distinctive corner of the field for themselves. Having canonized their work with reference to other scholars, the authors contribute their own episodic perspective to the subject at hand. This is ironic, for it enslaves a work to a tradition giving it legitimate existence and at the same time claims a certain novelty for itself. Rarely does anything new occur in journal science; I want to demonstrate this by showing the ironic literary production of authority within which arguments are embedded.

On the one hand, articles must be attached to an authorizing literature giving them a certain intellectual license. On the other hand, they must be sufficiently innovative that they do not simply duplicate the accumulated scholarship at hand. Although on the surface this would seem not to be ironic, indeed the very epitome of normal science moving cumulatively from one small accomplishment to the next, I would argue that innovation is only apparent. Instead, most articles only repeat and thus reinforce the literatures authorizing them in the first place. This does not mean that articles exactly repeat what has been said before; instead, they add little of value to the development of the narrow literatures within which they work. They exist only to give their authors something to do: build careers.

In establishing the validity of a topic by referring to a given literature propelling the topic into being, the literature is validated too. Yet topics, to be legitimate, must stick so close to the conventional wisdom of the given literature that they lose all their sense, their reason for being. This is why science articles are frequently so trivial. They add next to nothing to the canons authorizing them in the first place. Instead of innovating intellectually, articles merely add value to extant canons—ironically negating the modicum of wisdom they seek to add to them. By having to ground themselves so deeply in the work of others, topics necessarily fail to say anything of their own. If their titles and topics are nominally different, the differences are trivial.

It is assumed that disciplined scholarship of this sort contributes to overall intellectual advance. Incrementalism has been the model of scientific accumulation since the Enlightenment. But this faith is unexamined. The doyens and acolytes of disciplines like to believe it because their work thus has value and meaning. Yet it is not necessarily true that thirty-five articles on the same subject add more insight than thirty, twenty, or even two. It is an open empirical question whether disciplined knowledge, slavish to established literatures authorizing it, really accumulates. My reading of journal science suggests otherwise. The narrowly differentiated topics deeply grounded in authorizing canons only proliferate; they do not cohere into an ordered whole affording fundamental insights about the world.

So I want to understand scientific trivialization not as a contingent product of wrong authorial choices nor suggest that authors simply make better, more fundamental choices about topics. Rather, trivialization is a necessary outcome of the ironic authorization process governing topic choice in the first place. Indeed, authors do not choose topics but are in fact chosen by them. Once ensconced in certain literatures, authors are led to pose problems emerging from within the logic of those literatures; as such, they are enslaved to the literatures promising to set them free to do independent intellectual work. This is ironic because science could be quite different, scrutinizing established canons for faulty assumptions and thus interrogating their very reasons for being. But workaday science does not pause to examine its own authorizing aegis; instead, established literatures are points of departure. They are cited ritualistically, and in kind they turn topic choice into a hollow ritual.

Literature reviews oppress. Instead of serving as dialogue partners for inquisitive authors jousting with established authority, they enslave authors with a sense of overbearing tradition. The string citation of unexplicated experts in the first section of journal articles is a deadening exercise. The names parade by without rhyme or reason. Authority is parenthetical; one generalizes about a literature to which one pays obeisance by putting names and the dates of their publication in parentheses. This is an example: (Jones 1979). Science cannot seem to do without these citations opening articles. Without them, they would read as essays, even as poetry. With them, they sound authoritative, grounded, stable; on the basis of tradition, they make their own small claim to exist.

> The human capital perspective of earnings determination (Mincer and Polachek 1974; 1978; Polachek 1975; 1979) explains differences between men's and women's earnings as a function of variation in the productive characteristics of workers.[1]
>
> Studies of urban bias have produced conflicting results; some provide modest support for this perspective (Lipton 1973, 1977, 1984; Bates 1981; Ellis 1984; Nolan and White 1984), while others challenge its basic tenets (Seers 1977; Byres 1979; Herring and Edwards 1983).[2]

As such, these passages are typical of literature reviews. They summarize research done by others. In the first case, simple description suffices to characterize a particular perspective dominant in the relevant literature. Names and dates of publication are cited. In their own paper, the authors use and evaluate the "human capital perspective." In the second case, citation serves to suggest overt disagreement within a literature. Six books or articles are cited on one side, three on the other. Presumably the authors will add their own voice to this debate.

> Although the concept of class is central to Marxian and conflict theories of crime (Colvin and Pauly 1983; Spitzer 1975; Greenberg 1977; Chambliss and Seidman 1971; Quinney 1977; Taylor, Walton, and Young 1973), this concept has never before been operationalized in self-report research (cf. Hagan and Albonetti 1982).[3]

After citing authoritative uses of a particular concept, "class," the authors suggest shortcomings in that approach, namely the lack of operationalization (measurement). That is their angle; they take an extant literature and subject it to measurement, a norm of science, as we shall see. Citation authorizes the topic in the first place; in all three passages, names are marshaled to display the topic's legitimacy. Other people have written about the same thing. In the second place, citation serves to differentiate an established literature from the particular analysis at hand. It is crucial to observe that this differentiation does not take the form of opposition but rather of elaboration. The operationalization of the concept promised in the last passage does not overthrow the concept, and thus the tradition within which it emerges, but strengthens it. That much is not said. Yet it is clear that authority is not being overturned, only extended, by being subjected to precise mathematical measurement—"operationalization," as the term of art goes.

Consider this passage as a case in point.

> This paper extends the analysis of the determinants of urbanization of Firebaugh, Kentor and Timberlake by including their central independent variables—rural adversity and dependency—simultaneously in a series of equations. This procedure should permit us to determine the degree to which peripheral urban change is "a joint function of exogenetic and ontogenetic variables" (Snyder and Kick, 1979, p. 1101).[4]

The paper in question "extends" but does not overthrow the analysis mentioned. It apparently innovates by "including their central independent variables...in a series of equations." The paper promises little more than methodological innovation. Here science's literary irony presents itself forcefully. A topic must be canonized with reference to an extant body of knowledge; yet that canonization reduces its transformative impact on the tradition in question. In this case, the only transformation of knowledge is narrowly and trivially technical. In the prior case, a concept was "operationalized," not reformulated or scrapped altogether.

Now there is nothing inherent in the nature of science that requires this genuflection to authority and its subsequent delimitation of intellectual range and depth. Genuflection is merely a literary norm of science—what ordinary writers and readers take to be essential for an article to be authoritative. Thus science's text reproduces the norm subordinating it to repetition, thus only repeating it again. If there is nothing inherent in science requiring it to canonize itself with respect to a dominant tradition allowed to exert a powerful proscriptive force of its own in determining topic choice and thus reducing authorial autonomy, then the decision to write science this way must be explained in other terms. I understand it as a narrative practice, a conventional approach to rhetoric, and nothing more. Authors cite other authors in beginning their writing because others do so. Citation seems to unfold an argument from a solid beginning to a hesitant conclusion; within established authority, one presumably finds courage to add one's own small voice. But the literary effect of rote citation is ironically to subdue authorial voice in adding value to the already dominant canon. It is one thing to cite authority as a literary resource of a worldly, pluralist account; it gives others access to one's own intellectual autobiography—what one has read enroute to this particular formulation. It is another thing to rely on citation to establish the validity of one's topic in the first place.

Where citation aims to establish the legitimacy of one's approach, ironically it undercuts it. At least that is what happens in the journals. The papers I studied contributed next to nothing to knowledge; so obsessed were they with canon and method that they never assayed a distinctive point of view. There merely iterated what had been said before, perhaps at most rearranging the pertinent literature into a slightly new configuration or suggesting a methodologization of unoperationalized concepts. Of course, this does not make for scientific progress, especially when it virtually guarantees a narrowness of focus. Trivialization is imbedded in the authorization process itself. To stake a claim to space in journals, authors have to cite everyone under the sun in orienting their own small labor to the dominant force of tradition.

This suggests a way to think about citation that allows us to distinguish between different approaches to authority, tradition, canon. One may cite in order to leave a trace of one's own intellectual meanderings, recommending one's formative influences to others in pursuit of similar topics. The cited authorities do not weigh heavily on the argument but engage with it in lively counterpoint. Indeed, one does not have to agree with one's authorities; frequently that is one why writes in the first place—to protest and dispute. In the best sense, citation opens dialogues not only with readers who may read the same sources differently but with sources themselves. Even if dead and buried, authorities can be opened up through forceful imaginings about what they might have said here about current issues.

Unfortunately, this dialogic approach to citation does not prevail in contemporary science. Instead, authorities serve implicitly as evidence for one's own views. In citing them we try to mobilize the scientific community behind us. We acknowledge the normativeness of our views and suggest the legitimacy of the particular topics we have chosen for our work. This is not to deny the ideal of a community of science able to adjudicate arguments by appealing to a universal reason. It is to doubt that such a community exists today. Instead, in its place, we have disciplines, monologic and hierarchic organizations of knowledge in which authority flows from the top down and from the past forward. John Stuart Mill's golden ideal notwithstanding, the marketplace of ideas is as imperfect as the economic market. Objectivity does not prevail in the way scholars evaluate the work of others. Instead, these readings are driven by passionate self-interest and political perspective. Even if possible, Mill's undistorted marketplace of ideas surely does not exist today. The community of science is governed by conventionalist standards having a certain power behind them.

Psychoanalytic theory might suggest that we privilege established literatures simply out of the neurotic motive to belong and identify. Perhaps. In any case, established literatures have political force where they qualify and disqualify various intellectual (and hence social and political) approaches. The establishment of knowledge is not carried out in a vacuum; instead, disciplines are controlled by people with an interest in a certain conception of the world, and thus a certain world (to be brought about by the publication of their work). As such, literatures accumulate hierarchically. Some people have more control of them than others. Thus, when a writer seeks to establish legitimacy for a topic, she or he must wrestle with the dominant tradition canonizing topics in the first place. Whatever one may think of established knowledge at the time, one must pay it deference simply because it exists legitimately. It is inappropriate to ignore it if one wants to be given disciplinary warrant.

Discipline is circular. Resting on received wisdom, discipline is a norm.[5] It is what it has come to be. Yet it is more than that, having a powerful authority of its own. Discipline empowers and disempowers. The marketplace of ideas—intellectual disputation—is not open; it is constrained by the hierarchical organization of knowledge at the moment. The evaluation of intellectual claims trades on authority already established. Thus the discipline (here, science journals), in making its dominant intellectual consensus a norm, only reproduces its dominance. By requiring writers to acknowledge its authority, however ill deserved, discipline operates narratively to produce diffidence and reduce dissidence. Only an intellectual fanatic would send work to a science journal not already larded with literature citations establishing the narrow credibility of the topic at hand. Even with ample references and notes, it is not enough to dismiss them all as wrong. They must be obeyed at least to the extent to which they can be read to suggest our own legitimate intellectual topics. Undisciplined scholarship is an oxymoron today. Science proceeds canonically.

Resolving the Literary Irony of Science

To write science, then, one must root one's work in a tradition and yet prise free the possibility of one's own intellectual innovation within that tradition. I contend that this almost never happens simply because tradition usually has the last word; the seeming innovations made by particular authors are narrowly trivial at best. Many working scientists recognize this. Thus they propose research that does not contest with the tradition at all but simply refines it methodologically, enhancing the science aura of their work—another problem in its own right. The passage quoted above about the operationalization of the concept of class in self-report research is an example of this. Here is another one.

> Most current theorizing about corporate crime is too broad to be testable. We cannot assume etiological invariance between and within corporate crime categories. As Box (1983, pp. 19-23) points out, these crimes are extremely complex.[6]

The purpose of science is simply methodological refinement. If new knowledge cannot hope substantially to displace old, then at least we can develop better techniques for "testing" hypothetical claims. The vast majority of science articles are taken up not with concept development but with the improvement of methodological tools. Of course, this is to be expected in an age of mathematics and technical rationality. Progress is virtually equivalent to advances in our ability to manipulate symbols, figure, number; the computer is the vehicle of this advance. Thus, where science cannot really innovate, it hopes merely to improve its methodology, the topic of my next chapter.

> Much of the empirical literature is methodologically flawed... Most studies employ static models of variable relationships, isolating only a few years of illegal activity.[7]
>
> To date, then, Bourdieu's ideas have not been tested using multivariate analyses that allow assessment of the direct and indirect effects implied by his theory.[8]
>
> Finally, and most fundamentally, Easterlin's inspection of the data is plagued by a half-table fallacy.[9]
>
> Of course, Bonger had neither the data nor the analytic techniques to test adequately his prediction of the interactive effect of class and gender on delinquency.[10]

This is a way around science's literary irony. If fealty to canon is stifling, threatening to result in sheer repetition of established bodies of knowledge, authorial purchase can always be found in methodological advances, particularly the promise that theories will be "tested." In this way, literature reviews frequently lead directly into methodological discussions (usually the next section of science articles), consuming much of the body of the paper in question. Some literature reviews are more audacious than others, promising significant advances in the state of disciplined knowledge. Most, though, have more modest aims—test a theory, refine a technical operation, tentatively assay a new hypothesis or two. In any event, science possesses literary techniques to solve its own literary irony. When it is closed off from fundamental insight, it invents an agenda of reasonable intellectual aims.

> Although most such theories attach great importance to class, there is doubt about the correlation on which they rest.[11]

Iterative Citation

Citation as a literary practice occupies much of the effort of literature reviews. After all, one cannot simply list names of significant scholars on a topic; that happens in the bibliography appended to the article at the end. Prose must interweave the serial citations of experts and suggest a narrow path through the icepack to be taken by one's own work. In general, as I have suggested, the literary effort of the author is to portray his or her indebtedness to an established body of knowledge in the field in a way that leaves room to move; the author must value the existing literature and yet make way for his or her own voice. This ironic tension inhabits virtually all literature reviews. It is frequently resolved in the next section of the paper on methodology wherein one stakes a technical claim for one's own innovation.

In the meantime, before the methods section, authority is piled on authority as one establishes one's competence as a disciplined scholar.

> There have been a number of treatments of the growth of popular support for democracy in West Germany since the end of World War II (e.g., Verba 1965; Boynton and Loewnberg 1973, 1974; Conradt 1974, 1980, 1981;... Greiffenhagen 1984; Noelle-Neumann and Piel, eds. 1983), and numerous studies of generational change in West German values and political orientations (often in a comparative perspective: e.g., Jennings 1976; Allerbeck 1976, 1977; Inglehart 1977; Baker 1978; Barnes and Kaase, et al. 1979; Baker, Dalton, and Hildebrandt 1981).[12]
>
> We address herein the distinction between symmetric and asymmetric relationships in the context of a recent debate on the relationship between age and crime (see Hirschi and Gottfredson 1983, 1985; Greenberg 1985).[13]
>
> The effects of the size and structure of family of orientation on school achievement have been an enduring concern of social stratification research (e.g., Duncan 1965, 1967; Blau and Duncan 1967; Featherman and Hauser 1978; Hauser and Featherman 1976; Mare 1980; 1981; Sewell et al. 1980).[14]
>
> Finally,
>
> Each of our main findings also can be related to prior research and theory.[15]

The citation of authority never seems to quit. It is extremely difficult even to read the sentences larded with parenthetical citations as sentences; when broken up with long parenthetical strings of names and dates, they lose their direction. They repose on the page more as data, as evidence of the validity of one's claims, than as claims themselves. Only the last passage reads discursively (although it could as well come from the last "discussion" section of the paper). At issue is not simply bad writing. Of course, science eschews poetic possibilities. It wants to reduce the scope of authorial temperament by making the sentences read like unconstructed reports of authority, findings, description. The less elegantly the sentences read as careful prose constructions, the more readers will be convinced they are in the presence of science. Impatient to get their results across, science sentences present themselves virtually unconstructed, unrevised, unpolished—even if this belies the labor that went into them.

The momentum of the sentences is what matters in the literature review section of the paper. Once authority is cited and then compounded with the citation of other authorities, the reader virtually jumps from construction to construction, missing small prose fragments in a rush to get to a more comprehensible, down-to-earth section of the paper, usually the next section on methodology. After all, for practitioners already familiar with the names and dates enclosed in parentheses, nothing new is going on. We appreciate the nature of the literary ritual being undertaken: The authors are seeking authority on which to rest their topic. Important theoretical reinterpretation is not taking place; authorities are only being marshaled to warrant the rest of the paper, defending its specificity by providing a canonical context within which it can be read. The reader has only to glance at a few of the names to recognize the sense and sentience of the argument. The citations ensure a certain familiarity at the beginning.

It is also important to note the nearly universal absence of page references in these parenthetical citations. Some journals restrict page citations to instances where actual work is being quoted from another source. But depaginated citation has rapidly become the norm in science. It is enough to compact author and publication date into serial citations; the joined force of these "string citations" is the intended literary outcome, not a fine-grained explication of literature hinging on nuanced interpretation and thus the possibility of disputation. Of course, depaginated citation takes enormous literary liberties. One can compact almost anything and anyone into these sausages encased by parentheses; it is enough only to give a last name and state a year of publication. What matters is the claim one makes that these people support an argument or analytic tradition, not the validity of the claim itself or the nuances differentiating one citation from another.

It is one thing quickly to refer to a name in one's writing and cite only his or her book or article in question when one is trying to suggest some general affinities or simply to describe another's work. It is altogether another thing to depaginate a reference to another person when one is trying to claim that person on behalf of one's narrow argument. When one cites disparate depaginated authorities one is even on shakier ground. But the depagination of mainstream science simply shows the literariness of the apparatus of citation; one names names in an arbitrary way, crafting them into allies when alliances might be false on the evidence. The accumulation of knowledge is facilitated by the accumulation of citations.

Literature reviews meander through the thickets of names clogging them with parenthetical sense. Not gracefully styled as essays, the literature reviews with which most science papers open merely bide time, awaiting the more important work of methodological discussion and the description of results. One could as well dispense with literature reviews were it not for the disciplinary requirements of self-authorization. One must demonstrate one's disciplinary membership in authorizing the particular work one contributes to that tradition. The literature reviews disdain their literariness; instead, they are glorified bibliographies, mere iterations of names and dates. Others' work is simply noted, not discussed except in the most routine ways.

> The tie reproduction process has not been analyzed. While some assert that ties have a history (Mintz and Schwartz 1985), theories about the determinants of the interlock network's structure focus only on the initiation stage (Allen 1974; Pfeffer and Salancik 1978; Burt 1979; Pennings 1980; but see Galaskiewicz and Wasserman 1981).[16]

The last parenthetical clause, beginning with "but see," is a dead giveaway. There seemed to be at least a modicum of intellectual counterpoint in the serial citations listed in the passage. But the authors, in setting up the counterpoint, cannot quit equivocating. It does not really matter who said what; what matters is that virtually every prose word in the passage is matched by a name or date. The text speaks through its figural gestures, its iteration not of others' arguments but simply of their names strung together to create the impression of authority.

> Studies of urban bias have produced conflicting results; some provide modest support for this perspective (Lipton 1973, 1977, 1984; Bates 1981; Ellis 1984; Nolan and White 1984) while others challenge its basic tenets (Seers 1977; Byres 1979; Herring and Edwards 1983).[17]

At least there is the appearance of counterpoint. Yet the reader craves some elaboration of the names cited. In any event, the rhetoric of science echoes loudly here. Work is not argument intending to persuade but "studies" that "have produced...results." Work "provide[s] modest support" or does not. When the authors may lack sufficient data on which to rest their modest claims, they use other writers as data—indeed, as reporters of data, "results." Science not only deauthorizes itself into subjectless, authorless representation, but it deauthorizes other people. They are not given the right to speak for themselves when their words are not quoted and they are not even paginated. Seers is simply "1977," and he or she is made to share company with Byres and Herring and Edwards.

I am in favor of poetic license, conjuring an opaque world so that it is nearly transparent, or at least seems so. But the license used in authoring a world must be confessed methodologically, even celebrated, as the essay nature of science. These science writers I am quoting refuse their literariness while they take enormous license with others' work. They pretend unequivocal readings when such things are impossible. They have to argue for their interpretations, juxtapositions, and counterpositions in a way that suggests that those very accounts could be different. Unless they do, they are monologic, undemocratic. Science must compose itself in a way that the world could appear differently; in suggesting the playfulness and literariness of its account, it risks the possibility that it, too, will be believed. But, instead, science refuses its authoriality while continuing to author the world, the deepest duplicity of all.

I am not interested in opening up each of the sources just cited—Lipton, Bates, Byres—in reading them a different way. For all I know, the authors have done them justice. But in citing them iteratively as dumb evidence, the authors make themselves and their readers dumb. More has to be said than is said by stuffing diverse names and dates in serial parenthetical citations; if discussions of them are not offered, or passages quoted, or pages cited, then at least they should be acknowledged merely as embellishments, not the foundation of scientific verisimilitude. As such, depaginated, unargued literature citations merely clutter the paper with figure, a key rhetorical strategy of science, as we shall see.

Two things are happening here. First, others are cited as indisputable authority, virtually as data themselves. They thus prop one's own argument as well as provide a canonical context within which one may proceed in specifying one's own topic. Second, and less obvious to a linear interpretation, the parenthetical citation of serial sources becomes an obdurate text of its own; the weight of authority seems self-evident when presented this way. And writing crammed with these citations has less time and space for words. Like the use of figure and number (discussed in the next two chapters), citation both diverts the reader from the words themselves and imparts a scientific aura to the text. To the extent to which serial citation has become a norm of science, it is mistaken for its essence. Writers cite and readers expect citations.

The issue is not whether citation per se is good or bad. One would correctly conclude that that depends on the text and topic at hand. If citations are used, however, they should be as dialogical as possible; that is, they should refer to specific work that can be checked, and they should be explored discursively. As mere buttress, citations defeat the notion of science as a communitarian language game, a notion I argue for later in the book. Reference to other authors does not prove anything; at most, it demonstrates the existence of a common problem approached differently by people. When I write, I often refer to people working in the same area but who either formulate the problem differently or answer it differently. I cite them either to show a common problem approached differently by different people or simply to engage with wrongheaded writers whose work I take seriously. Sometimes citation, for me, serves the purpose of illustration especially when I suggest an argument but do not or cannot follow it further. Then I might cite work that extends and illustrates my argument.

Citation must exist in counterpoint to one's own approach, even if the differences appear friendly. If others simply "support" one's own "findings," then one's work is useless. The same paper rewritten twenty-five times is still the same paper. But positivist science proceeds accumulatively, collecting experts in defense of a common conception and elaboration of the world. But this is a deceit when, in fact, the same paper cannot be simply iterated over and over; each version changes it, producing a new account itself subject to reformulation. Positivism as a literary technique compresses and thus reduces differences between authors in the interest of intellectual hegemony. In homogenizing real differences it supports its contention of a common worldview and thus a common world. While agreements are certainly possible—ever the aim of dialogue, as Habermas profoundly argues—(that citation of Habermas being an example of a passing moment of consensus), they are essentially ephemeral. One may touch other people in one's narrative approach to the world, even moving the world forward through that account and the community it creates, yet one will still be fundamentally alone. Texts are as differentiated as the people they express; consensus only indicates that we can be alone together in those moments when we share a fundamental worldview and thus inhabit a common world peacably, the aim of all good politics. Agreeing just to agree is not the aim of writing. Writing would change the world to which it is an address.

Citation does not strengthen one's argument but only builds a common community out of which nonantagonistic differences can spring heterogeneously. I understand a version of socialism to be simply the facilitation of difference within sameness; as a language game, socialism allows consensus, indeed requires it, yet it withers on the vine if consensus takes on a life of its own. Canons reify consensus when they suppress the possibilities of their own interrogation, even their right to exist. This technical discussion of citation opens up larger political and metaphysical discussions of the nature of authority. Consensus solves no intellectual problem but simply provides a context within which certain social relations of truth play themselves out. Consensus dissolves into monologue, and thus oppression, when the truth is iterated from author to author, generation to generation. Consensus refreshes itself only with dialogue—writing's openness to its own corrigibile finity, its sense that it could have been done differently. Writing opening to its own animating desire opens itself to other versions, thus creating community. Writing that merely views writings as pieces of intellectual capital to be accumulated and then horded disciplines itself; after all, disciplines are only hierarchical organizations of knowledge, privileging some accounts and disqualifying others.

Science indebts itself to what it calls "a literature." In this, it refuses its own literariness. For science, literature is the vouchsafed body of formulations, findings, recipes on which everyday science must build. It is no accident that this is precisely the model of capital accumulation. Like a private economic order concentrating value, cumulative science centers and concentrates truth. Power becomes a norm when people accept it unthinkingly. Thus readers learn how to write by reading everyday science. The literary tricks worked by science are picked up by science readers and thus reproduced. Serial citation becomes essential to science's text; its contingent normativeness—merely a particular rhetorical strategy preferred over others—is learned and taught as the nature of scientific investigation. Authorities proliferate and thus accumulate into a disciplinary canon, a literature. But citation is literary work that can be accomplished in any number of ways.

Science's basic mistake is to believe that citation in itself decides epistemological questions. It does not. Those questions are only resolved in disputes that necessarily proceed groundlessly; there is no Archimedean point from which we can resolve the babel of tongues. I prefer to read and thus write science as a ground of community; truth is only a way of formulating the aim of community differently. In the social world, truth is the emancipation from canons weighing too heavily on every literary endeavor. Positivism disciplines by concealing its literary nature. Anyone can conjure the world. The rhetorical devices of science decide nothing, but only signify obedience to science's prevailing norms. But the norms could be different. We could decide that science must not cite but only essay itself; or that citation must be quoted and paginated; or that citation must always exist in counterpoint to one's argument. Anything is possible in the literary way of science. Norms are made to be broken.

But I am saying that the norms are not accidents; today they exist to discipline defiance. One cannot earn space in disciplinary organs without demonstrating disciplinary competence—how to formulate and figure page 1 (requiring one to be employed as a disciplinary worker in the first place), how to review literature, how to figure findings, and finally how to "discuss" them. In deauthorizing itself, science authorizes authority. Ironically beholden to a literature delimiting it in giving it the license to write, science kowtows to a pregiven discipline—authorities, norms of presentation, methods, findings. This is appropriate if we accept the right of separate disciplines to exist; I have said that disciplines possess no inalienable rights.[18] I say this to make a political point. Today disciplines discipline the unruly rebels who write what is dismissed as non-science—polemic, politics, poetry. The apparatus of citation plays a role in this disciplining. It grounds writing in a canon defining the various subliteratures of the discipline.

This is a problem from the point of view of intellectual innovation. Deference to authority is a problem only if the authority is illegitimate. In spite of itself, science fails to resolve its own literary irony. It is swallowed in the literature it swallows; journal articles are trivial by definition. They exist too close to the literatures authorizing them. Thus they deviate too little. While the possible configuration of citations is infinite at the margin, in fact people within disciplines reproduce the same sources, in effect rewriting the same article endlessly. One may not cite anyone in any fashion. Certain canons dominate and thus they constrain the intellectual work done in their name. As disciplines mature, adding a larger stock of names to their bibliographical repository, new names are added to old. While there may be slow intellectual shifts, the basic substratum of disciplines does not change. Citations reproduce themselves; that is their nature. In authorizing one published account, they tend to authorize others done in its name. Articles are connected only in the sense that they add themselves to a certain preexisting stock of knowledge—the literature. One will not be published if one avoids the standard citations. Indeed, one merely hopes to add one's own name to those citations.

People who work in science disciplines understand this well. One's esteem is a result of one's integration into the disciplinary literature. University committees even measure this by counting one's citations in the various indexes of natural and social science; the more the better. We assume that citation is good, indicating intellectual merit. When the quantity of one's science publications is one measure of intellectual quality, one's number of citations is even a better measure of disciplinary stature. Publishing in obscure journals is devalued as a result. Of course, this is circular: The more obediently one cites the dominant literature in a particular intellectual area, the greater are the chances that one's work will be published by the major journals. And thus the more value one adds to this dominant literature, inherently devaluing work falling somehow beyond the perimeter of dominant authority.

Making Science Claims

While obedient to canonical authority, as they must be, science articles must also stake their own claims. It is not enough to appear only to summarize extant literatures; one must appear to use these as points of departure for one's own particular contributions to their overall evolution even if, as I contend, this appearance is typically fraudulent. Literature reviews contain or end with authorial claims to intellectual distinctiveness. They promise to make their mark within canonical context.

> Despite the advances contributed to criminological theory by the Hirschi and Gottfredson article, Greenberg finds many of the authors' contentions to be overstated and misleading... To address the symmetry question, we begin... Then we...[19]
>
> Yet we know of no empirical studies examining whether men who have female workmates actually evaluate their situations less positively than comparable men in male-dominated work settings.[20]
>
> Thinking of schoolwork as work, we build on research that has shown that the opportunity to exercise self-direction in paid work has powerful effects on adult workers' orientations to self and society. We here investigate whether the opportunity to exercise self-direction in schoolwork similarly affects the personalities of adolescents and young adults.[21]
>
> Yet research in this area has approached a theoretical impasse from two different directions.[22]
>
> The literature contains few detailed hypotheses about the relative effects of agency and systems factors.[23]

Science warrants its claims to distinctiveness by "building on" past research. Or it "know[s] of no empirical studies examining." It would resolve "theoretical impasse[s]" and offer "detailed hypotheses." These claims allow the science article to shift into the methodology and findings sections. Not much in the way of novelty is promised; yet there is enough on which the authors can hang their hats. Science "builds on" past science; as such, it respects tradition and conceives its contribution as evolutionary, not revolutionary. Virtually no article will fail to include these sorts of promises. Without them, there can be no narrative transition between the literature review and the discussion of methodology in the following sections. Near the end of the literature review, authors impatient to get on with their analyses seek to make and then warrant their claims to journal space. Without authorizing their work, they fail to appear disciplined. But without the promise of minor innovation, discipline overwhelms them; they cannot move on to the particular analysis they have done.

It is useful to read literature reviews as the last part of science papers actually written;[24] I suspect that is often the case. Yet the logic of presentation suggests otherwise. First, it implies, scholars read many books and articles bearing on a particular problem. They then tease out a testable proposition or infer a methodological or conceptual lacuna with which to occupy themselves usefully. While this may be the way science occasionally works, in fact the literature reviews only suggest the autobiographical process of professionalization undergone by employed university scientists. One is immersed in literatures in graduate school and then one learns to exploit them for publishable work warranted by one's claims to create new knowledge. The literature review is such a warrant, even though in effect it describes work already done.

In reality, science contingently and often haphazardly stumbles onto its technical and conceptual problems in the process of doing certain work firmly rooted in tradition. Already assimilated, the canons cited in literature reviews are reconstructed after the fact to frame the particular problems taken up in everyday work. For many people, this is the hardest section of the paper to write because it forces them to go backward to graduate school and to their early academic careers when the authorizing literatures were more clearly in mind. In fact, frequently writers have not even read the work they cite but find it after the fact, reconstructing it to suggest a question to which they have already found the answer. Of course, there is nothing wrong with intellectual serendipity; I suspect it is the way many of us work. We are doing one thing and then confront a problem that excites us more; or we simply must solve the unforeseen problem before we can get back to our original work.

The problem with science's implied logic of discovery—first, read books; second, derive research topics; third, explore them—is that it suggests false continuity between one's topic and the authorizing literature.[25] Science wants to show that topics have been derived from a legitimate body of knowledge; otherwise they will be seen merely as poetic choices made for no other reason. Topic choice is reconstructed in a disciplined way in order to mislead readers that that is the way the analysis really came about. And that only redoubles the value of the authorizing literature implied to have impelled the analysis. As well, it bonds the topic to literature in a way appearing to suggest more continuity than may really (or should) exist. This problem looms especially large in chapter 8, where I discuss the way science "discusses" its findings in its summary section.

Topics are forced to fit a disciplinary mold, when in fact their gestation may proceed outside the authorizing literatures they cite. In fact, they may devolve from inspirations quite outside the discipline they claim for themselves. The norm becomes necessity; intellectual work must derive itself and thus it becomes derivative. This is not an argument for atheoretical intellectual work. The contingent ways one comes to problems can always be reconstructed out of the complex weave of biography, authority, history. Problems never simply appear on one's doorstep. But this reconstruction threatens disciplinary power where it involves a reconstitution of authorial subjectivity (what I have read, where I have been, my political desire) believed to be antithetical to positive science. And it may call on nondisciplinary, even undisciplined, intellectual sources threatening a discipline's authority structure.

I do not mean to read trivial work as more global and thus more important than it is really is. Much science remains too constrained by its own subordination to canon and method, especially the figural fixations of mathematics. Yet I want to understand this constraint as a narrative outcome, an outcome of fitting a literature to a topic conceived and investigated independently. Every science writer knows how inadvisable it is to submit work without citations; similarly, if one were authentically to reconstruct the complex lineage of a topic, one would be refused publication. Thus writers scramble to anchor their work in disciplinary literatures seeming to bear fit to their particular methodological and conceptual departures.

In part, then, the trivialization of workaday science comes about largely in response to the requirements of literature reviews. Perhaps particular analyses treat of more global issues than they are allowed to say. At least, their narrow treatments may not stem from literatures but only enhance their value once reconstructed as such. And I am exaggerating the independence of everyday science from authorizing literatures; working scientists keep an eye on the literary auspices under which their surprising, meandering, accidental work might be seen to proceed. After all, they realize that they must reconstruct their twisted intellectual paths to a topic in a literature review that they write last but publish first. Authors recognize their own literary responsibility for ensuring continuity between their theory, methods, findings; yet they fail to learn from this about the essentially fanciful—undisciplined—nature of science.

Authority trivializes science by dictating problems to it. Authors trivialize science by hinging it to authorizing bodies of knowledge. Both are true. The author is not entirely out of control of the compass of the work in question. It need not be derivative; yet it must read derivatively for it to earn a disciplinary imprimatur. In any case, trivialization is not the only problem. Authors also add value to existing canons by tracing their topics to these canons deductively when in fact there may be real discontinuity between them. As such, then, they reproduce particular canons by citing them. And the very requirement of authorization in the first place virtually guarantees that topic will never liberate itself from its constraining disciplinary context.

The delight and mystery of science is that it reveals a region of the unknown it only partly illuminates. It genuinely surprises itself when it digs beneath the surface of its chosen topics. A little knowledge gained only reveals much more to be acquired. Science always revolutionizes itself by undercutting its original formulation of the problem; it discovers better formulations of the problem, better method, better conceptual apparatuses—even better interpretations. A better way to write science might involve the naturalistic accounting of the logic of discovery. One would shift between the engaging topic and an account of how and why the topic engaged the writer in the first place, entering autobiography, aesthetic curiosity, and political desire directly into the text. Today all these things are suppressed in the pretense that science devolves linearly from existing literatures, adding value to those literatures and making science a slave of acknowledged authority. At some level, science is always free to constitute the world, thus contributing to the creation of that world. Discipline narratively requires authorial suppression of the real logic of inquiry.

Let me offer a tentative example. At least, let me spell out where reconstructive work may begin.

> Unfortunately, no single cross-national study has simultaneously tested the hypotheses advanced by these three theories, though one paper has tested modernization and dependency theories (Kentor 1981) and another has tested urban bias and dependency theories (Bradshaw 1985b).[26]

A cursory reading would reinforce the impression that there are three—and only three—major theories in the subfield in question. One would also gather that these theories imply hypotheses that must be tested cross-nationally. Finally, one assumes that it is worthwhile to test hypotheses as a way of evaluating competing theoretical claims. Too much is taken for granted. The author is trying to stake a claim to his intellectual distinctiveness by promising what he says other writers do not deliver— comprehensive hypothesis testing using a single cross-national study.

I doubt that the study was undertaken exactly for that reason. Indeed, I suspect that many people would disagree that the field is covered by three theories. And one might argue with the implication that theories can or should be tested. While the passage just quoted seems to warrant the author's claim to be doing useful science, it leaves too much unexamined and fails to disclose its own literary gestation. Let me take another example.

> Of course, Bonger had neither the data nor the analytic techniques to test adequately his prediction of the interactive effect of class and gender on delinquency.[27]

Who is this Bonger? Is he important? Why is he important to the author? Did Bonger either want or need data and techniques with which to test his prediction? Did he in fact make predictions? Should we? Did the authors come to their problem through Bonger, because of Bonger, or in spite of him? Perhaps it does not matter; yet they are the ones warranting their work by promising to improve on Bonger, just as the previous writer promised a definitive test of some competing or complementary theories. While it may be useful to contextualize one's work in the light of its intellectual antecedents, this must be an accurate reconstruction really describing the topic's genesis. It is not that I doubt the particular claims advanced; I merely suspect that they were appended after the fact, after the research, in aid of narrative legitimacy. After all, who is this Bonger? And are there really only three theories bearing on the problem above?

Science scripts too much continuity into its account of its relationship to authority. Even if it does not stand authority on its head, it may frequently just ignore it, coming to problems and then resolving them in undisciplined ways. The empirical existence of disciplines constrains science writing by requiring of it a disingenuous account of its own logic of inquiry. Working scientists impatient with these phonied literature reviews write them ritualistically—to get published. It is a positivist assumption that piecemeal research must demonstrate its contribution to the ordered advance of the whole. But what if there is no ordered advance? What if writing does not make a contribution, perhaps even derails disciplinary accumulation? These appear not to be possibilities for mainstream science. Literature reviews replace a literary reflexivity about the genesis of one's own work in the context of real lifeworlds comprised of autobiography, history, politics.

> Five of the most eclectic contemporary theories serve as the principal source for the composite model developed in this paper.[28]

> Several major research traditions suggest the hypothesis that is fundamental to the present inquiry—*that the opportunity to exercise self-direction in schoolwork influences noncognitive*

> *as well as cognitive aspects of students' personalities.* Three bodies of educational research are particularly pertinent.[29]

When many authors tortuously reconstruct their logic of inquiry deductively, allegedly deriving neatly from a given body of knowledge or theoretical orientation, others, like the ones, quoted aboved, claim ecumenism when it comes to the tenuous relationship between their topic and an authorizing literature. Of course, another reason why they celebrate their seeming eclecticism is simply to enhance the authority of their analyses. The more "major research traditions" or "contemporary theories" they can claim in support of their work, the better. This is the atypical strategy; the more common approach is to suggest some counterpoint or controversy in the field and then promise to shed light on it in the subsequent empirical analysis. In each case, though, the literature review only weakly structures the analysis to follow. In a certain sense, it is quite gratuitous.

Science claims promise the advancement of knowledge. That is why people take the liberty to offer their work to the scientific community. But while much published work is interreferential, there is a sense in reading science articles that their incredibly high degree of specialization, along with their tenuous grounding in a given literature, makes clear-cut disciplinary advancement problematic. After all, few really follow the path blazed by particular journal authors. It appears that many articles simply fall into a black hole of sorts; while they may earn a few citations over the next few years, they do not serve to settle debates in the field or forge new paths in conceptualization, theory, or methodology. They merely advance careers; simply occupying the expensive real estate of the leading journals is enough to earn tenure, promotion, merit raises.

In this sense, the notion that science accumulates in a reasonably continuous way, each article grounded in its authorizing literature subsequently expanding and improving that literature, is a regulative idea. Science writers would like to believe it because it gives their work a reason for being apart from sheer professional self-aggrandizement. Sometimes articles are so widely cited that they do make a dent in the conventional wisdom. But this is rare. In the 1940s and 1950s, articles could have had greater impact if only because there were fewer writers, readers, and articles. Davis and Moore's notorious 1945 *ASR* article defending inequality on functionalist grounds is a case in point.[30] In fact, it initiated a whole debate breathing life into a stagnant tradition of American stratification research.

Today the sheer number of science articles is too high to make general scholarship possible. Indeed, even specialized scholarship must restrict itself to works just like itself. Ironically, few works resemble one another;

each topic seems virtually sui generis even to insiders, people already familiar with many of the references and methodology. The notion of a cumulative body of knowledge moving slowly but surely forward is a myth justifying specialized, self-referencing work. As long as people can prop their idiosyncratic analyses with a preceding literature review, they are given a hearing. This is not to imply that a monolithic, perhaps state-controlled science would be better; conformity is not an antidote to deceptive idiosyncracy and self-trivialization.

Literature reviews have become simply talismans, suggesting the pedigree of an analysis. If one owns or knows certain books and articles, one can fashion a literature review post hoc to legitimate the analysis already carried out. The important thing is that science appears grounded in a wider universe of discourse among people working on the same problem and together advancing the state of collective knowledge. But people cannot and do not work in concert when the object of their inquiries has so much specificity that it is almost entirely an artifact of a certain literary approach to a topic—title, references, methods, figure, analysis. People cluster together in various subfields within science disciplines for shelter and succor; my own discipline proliferates these "sections" devoted to environmental sociology, theory, the sociology of emotions, family, criminology, and so on. These are the closest we can come to communities within highly differentiated and bureaucratized disciplines. We have little in common intellectually apart from knowing some of the same references and having a certain literary facility in the way we conjure the world in our science. That is to say, we share science as a literary norm but fail to embrace a common intellectual milieu within which we can make sense of one anothers' scribblings. All we can do is meet (and promote ourselves) at annual professional conferences. No one reads the journals anymore, but everyone writes for them. They make no sense.

Chapter 5 Science Aura I: Method as Topic

Science Aura

Science proceeds on the basis of citation and figure; through them it enhances the authority and representationality of its analysis. In chapter 4 I discussed how various uses of citation in papers' opening literature-review sections enhanced the appearance of their authority. In this and the next chapter I want to examine the ways in which the use of figure and number enhance the representationality of the text, the impression on the part of readers that the world is virtually present in the paper.[1] Whereas number is a particular formulation of the text's general use of figural gestures, it also enhances the text's seeming authority;[2] the math effect adds value to the authorizing apparatus of citation. To open this discussion I want to address the nature of the science aura, especially the way it buttresses journal science at crucial points in its logic of presentation. In effect, science needs a science aura because it otherwise fails to make good on its promise to deliver the world whole—an impossible aim anyway.

The science aura is the embellishment of the text with design, figure, and number, which in effect become texts themselves. When the prose lacks a certain authority or there is trouble relating the larger promises of the literature review to the narrower outcomes of the particular analysis, figure steps in to win the day. Science redoubles the authority secured in the literature review by presenting its world in the ineffable but prestigious argots of tables, charts, equations, models, causal flows, correlations, and the other artwork of quantitative method. Indeed, it is important to realize that the issue here is not simply quantitative method because, as I shall show, so-called qualitative articles secure a science aura too. All sorts of science articles fall back on figural gestures to clinch the argument; the employment of math is only a local case of figural presentation, albeit by far the most common one in contemporary science. Even articles that are not quantitative gesture figurally in order to make good their claim to be science.

So I consider methodology as a literary topic, indeed as a narrative technique.[3] Too frequently, critics of positivism have attacked mathematics as the source of science's soullessness.[4] That is only trivially correct. Math is only a special case of the science aura gained by the proliferation of science's technical and figural embellishments in the midst of (increasingly in place of) words. Even without the use of statistical inference, science can gain an aura for itself, rebutting challenges and winning friends. Statistics is no more the essence of science than are coauthorship, abstracts, and literature reviews. Science figures itself in the broadest sense in order to convey the world's presuppositionless representation; math is only one literary strategy in this. I am not an opponent of methodology, quantitative analysis, or statistics. Even to formulate the problem in those terms accepts science's wrongheaded duality of hard and soft methods, quantitative/qualitative, empirical/theoretical, value-free/value-laden. All these dualisms misrepresent the marginal alternative as politically jaundiced and unrigorous. The real problem with positivism is a literary—hence political—one. Positivism believes that it can show the lawfulness of the world it pretends to portray without the aid of presuppositions, propositions, constructions. By doing so, it secretly hopes to complete a world it knows is incomplete; in the fashion of every ideology, it depicts a world without exit as a way of blocking the door.

Science uses citation to ground itself in an authoritative tradition. It uses figure abundantly to portray the present world, without mediation. As literary strategies, these act politically; science's world is only a fiction, a constitution of a driven author cleverly factoring his or her subjectivity out of the account. But just as literature reviews authorizing analyses are fanciful prose constructions, so figure and number are poetic gestures in their own right. Graphs, equations, tables do not clinch any argument; they merely figure it, suggesting indubitability of an analysis. This is the science aura—the literary production of presuppositionlessness. Another word for this is objectivity; by figuring and mathematizing the text, all else being equal, science convinces.

In what follows I trace the steps by which this aura is created. Here I want to indicate that methodology is at issue not as intellectual tools or techniques possessing an inherent essence (e.g., statistics as good or bad) but as the figural assemblage of prose, design, and number into a total text. Made up of thoroughly contingent fragments of description, display, and analysis, method reflects certain literary norms governing science's presentation. Methodology is a textual accomplishment, not an invariant apparatus through which data are dumbly processed.[5] As such, I want to deconstruct journal science as a fictive pastiche simply conjured out of the normative tools thought to make science science. That these tools do not make science science is part of my argument; there can be other narrative arrangements of citation, figure, number, and prose that portray the world not as an inert blob governed by laws but as the dialectically open process that it is (and might become).

I argue for a new concept and practice of science, drawing from the inspiration of various left-wing critics of positivism since the early twentieth century. When in his (1969) *An Essay on Liberation* Marcuse prophesies a "new science and technology" fulfilling the "play impulse," he wants not to destroy high science and technology as a latter-day Luddite but to reformulate the emancipatory possibilities of science and technology.[6] Here I want to argue against positivism for a new science, the possibility of which is available on literary grounds. Once we understand that positivism is not an invariant set of intellectual techniques performed on an unresisting world, we can then read it as a literary accomplishment, a particular way of arranging the page of science narratively in order to persuade.

This requires us to read science rhetorically, as a text intending to bring about a certain world by suggesting a metaphysical conception of that world. Science is argument; in the case of positivism, it argues through seemingly unmediated description.[7] Of course, no textual address to the world captures it fully or essentially. The page always stands at one remove from its object; this is its literary quality. Science conjures the world as representation. Whether or not this representation is accurate, science in representing also constitutes. The citations, figure, and number deployed in the text do not neatly correspond to fixed objects in the world. Citations construct cited texts as intellectual allies; figure depicts the world in a certain imagined shape; number counts what it implies are worthy of being iterated. Methodology reconstructs science's logic of literary presentation as a rhetorical strategy; it is a way of enhancing indubitability.

This is directly political when method, through ontology, seeks to portray a fixed world. Positivist science uses citation and figure to prop its representation of alleged social laws; by so doing, it hopes to disqualify defiance as impossible, irrational. Positivism is ideology in that it suggests a world to which conformity is the only reasonable response. While allowing some room for personal self-betterment, positivism today (including bourgeois economic theory and religion) suggests a frozen history.[8] Fundamentally, it postures, things must be the way they are—that is, the way they are represented in the lawful text. Thus positivism's political efficacy depends on its literary abilities to portray a single lawful universe; the science aura accrues to methodologically artful work. Science seeming scientific, according to the prevailing norms surrounding scientific presentation, willy-nilly wins the day.

Particular/General

The science aura must be cultivated. Articles open with literature reviews and then lurch into exhaustive descriptions of the methodology of analysis.

For many working scientists, this is where they begin reading. Authors describe their database, sampling techniques, the sorts of statistics employed, the nature of their research subjects; they also specify their equations—construct manipulable hypotheses in a multivariate form. Let me offer a few examples of the things discussed in this section of science articles:

> Logistic regression is necessary because the dependent variable (respondent's class) is nominal level and does not allow the application of OLS regression.[9]
>
> There are, of course, methodological limitations to newspaper-derived collective action data. As critics have pointed out (Molotch and Lester 1974; Danzger 1975; Snyder and Kelly 1977), news reportage does not perfectly "mirror" social reality.[10]
>
> To address these questions we drew a stratified random sample from the population of 3,674 full-length articles published in 10 major sociology journals in 1974–83.[11]
>
> This paper analyzes data from the *Quality of Employment Survey* to explore how gender segregation affects men's psychological orientations toward work.[12]
>
> To address these issues we have tabulated NORC General Social Survey (GenSoc) reports of church attendance for all years from 1975 to 1984. We analyze a cross-classification of attendance by age, cohort, religion, and sex. Year is excluded because it has no significant effect on Catholic or Protestant attendance after 1975 (as shown in the next section). The GenSoc attendance question differs from the Gallup question in time reference and number of alternatives offered to the respondent. "How often do you attend religious services? (0) Never, (1) less than once a year, (2) about once or twice a year, (3) several times a year, (4) about once a month, (5) two or three times a month, (6) nearly every week, (7) every week, or (8) several times a week?" We collapsed categories for our analysis: (0) "never," (1-3) "yearly," (4-5) "monthly", (6-8) "weekly." We take advantage of the absence of year effects to make fine distinctions among age groups and cohorts, giving us more precise estimates of these effects than were possible using the Gallup data (because we had to control for year in that analysis). The categories used are shown in subsequent tables. We began our analysis with a four-way table having 496 cells. It turns out that sex and religion interact with both age and cohort, so for ease of

> presentation, we treat the analysis as if we carried out a separate analysis for each of the four combinations of sex by religion.[13]

A number of things are obvious here. Whereas the literature reviews were conducted in somewhat generalizing language, referring to theories, even whole intellectual traditions, these discussions are extremely particularistic and fine-grained. The literature-review section abruptly shifts into this close discussion of the mechanics of science: "496 cells." This is a remarkable transition[14] already anticipated in the articles' solution to the problem of literary irony, namely that authors authorized their particular work with reference to whole traditions from which they derive in a tense relationship of subservience and independence. Science moves very rapidly from the general to the particular as modes of presentation. Whereas in the literature-review sections whole groups of authors are sausaged into parentheses in support of terse characterizations of large intellectual traditions, the methodology descriptions that follow are minute to the point of absurdity.

> For details on these and other sampling procedures see Davis and Smith (1985). At the time of the present study, the General Social Survey had been conducted in 12 differents years: 1972, 1973, 1974, 1975, 1976, 1977, 1978, 1980, 1982, 1983, 1984, and 1985.[15]

> Before we introduce measures of theoretical concepts, it is necessary to describe our data...The survey was conducted during the first four months of 1979...We paid each student $5.00 to participate at school in the survey.[16]

> To examine the determinants of interracial friendship stability, a dyadic-level analysis is required. In each dyad, P is designated the chooser and O the student who can be chosen. We examine those dyads in which P chooses O as Best Friend at some time during the course of the school year. Our interest is the stability of that choice. The dependent variable for the descriptive analysis in Table 2 is the termination of P's choice of O (Dissol), coded as unity if the friendship dissolved and zero if the friendship continued.[17]

The first passage is extraordinary in its serial listing of the dates after the colon. It is hard to imagine that we needed to know this; even if we did, surely the authors could have abbreviated like this: 1972-1978. But the point is both to include minute technical detail and the abundant numbering of the analysis in order to enhance the science aura. Just as names and their publication dates were packed solid into parenthetical citations in the literature review as a way of authorizing the analysis, so here detail and number proliferate as subtext.[18] In the second passage, we surely do not need to know that the study was done in the first four months of 1979. What if it had been done in 1978 or during the last four months of 1979? Again, detail prevails as telling text. As well, why should we know that the student subjects were paid $5 for their efforts? Do we care about the equity of their compensation? Perhaps we should. I suspect the point is to lard science with detail as a way of privileging its account.

The authors would doubtless defend themselves circularly. This is what people do in science; thus this is science. They fear that without close accounts of their method, they will be accused of shoddy science; the anxiety about fakery in science is ever-present today. What is interesting is the rhetorical shift from issues of high theory in the first section of science papers to mundane matters of method, sample, statistic in the second section. Of course, nothing requires papers not to proceed in this way; the rhetorical shift from general to particular is interesting as a fundamental lacuna in science, revealing a telling discontinuity between the seeming theoretical concerns of literature review and the particular analysis actually preoccupying the bulk of the article. The rhetorical shift from general to particular would seem to confirm my sense of science's labored reconstruction of an authorizing, disciplining literature that in fact is added to the finished article post hoc. Science really wants to tell of the small details constituting its daily work—test tubes, datatapes, computer runs, scatterplots, regression equations.

As a text of sorts, the microscopic methodological detail of science articles both enhances their science aura—the overall project of positivism—and gives poetic vent to the repressed author. After disciplining himself or herself in the literature review, the soul of the scientist breaks free in the fine-grained reportage of the next sections. It is nearly impossible for humanists to recognize poetry in number and figure. I am not simply saying that they are beautiful; that is not the issue. But scientists love the detail of their work in the same way that Shakespeare scholars love his sonnets and worry about the meanings of the Moor. It is hard to separate norm from passion here. Are the articles larded with detail simply to enhance the paper's scientificity or to express the poetry of discovery? My answer is, *both.* After all, scientists select themselves to do science. They are the ones who love tinkering with number and figure; a colleague of mine said that he "just loves to solve the intellectual puzzles" posed by the topics of journal articles. He does not do science because he wants to contribute to the literature authorizing him but simply to have fun with the only intellectual work he knows. Of course, he is not oblivious to the career benefits of mainstream science scholarship. But there is still passion in the path analyses for him.

Yet just because something is art does not mean it is beautiful or good. People are not born to be tinkerers; they learn it. When scholarship is diverted in the crabbed routines of daily science with its overwhelming methodological preoccupation, something is wrong. Although daily science, like any expressive endeavor, is governed by certain folkways that become important to it, these are not essential; science could conduct itself differently. Scientism is the belief that science can solve all social problems; everything is corrupt except positivism's methodological apparatus. As Adorno and Horkheimer eloquently argued in *Dialectic of Enlightenment* (1972), science becomes scientism—ideology—when it exempts only itself from critical scrutiny of its metaphysical, hence political, desire. The belief that science avoids metaphysical investments is the deepest metaphysic of all, thus the hardest to uproot.

When I say that science is poetry for scientists, I mean that they temperamentally resist the portentous solemnities of the literature review.[19] They are trained to revel in the minute discussions of method taken to be science's essence today. But the political problem is that science is not essentialized in statistics, sampling, or regression analysis any more than philosophy is exhausted by modes of presentation like that of Plato's *Republic*. Indeed, one of my main points is that science is poetic and philosophic and that poetry can be a kind of science. I refuse to give up the notion and practice of science to the positivists, who reduce it to the folkways of mainstream methodology. If science is simply an orientation to the object world, then I am a scientist and want all of us to be scientists as a vital requirement of the good political community. Democratized, science checks tyranny. A kind of fiction, science is unique only in that it must claim validity for itself; where Graham Greene knows that we know he intends to portray the world truthfully, we must convince readers of science's truth. How we do this is up for grabs. I do not rule out the mathematizing rhetorics of dominant journal science but only suggest that there are other ways to make validity claims stick.[20]

From Theory to Science: Method as a Literary Logic

I read methodology narratively. Like any fiction conjuring a world out of semiotic units, methodology is the reconstruction of science's chosen logic of inquiry. One does not need to use advanced methodologies of statistical inference and causal modeling in one's one work to write about it as literature. One does not need credentials to be a scientist anymore than one needs diplomas to read and write philosophy or poetry. Scientism suggests otherwise; it arrogates to itself the right to disqualify rhetorics as non-science—hence nonsense. But we can crack the codes, demystifying them and opening them to other writers. I am less interested in the technical vagaries and possibilities of mathematical investigation than in the way it sits on the page, crowding out prose and producing the science aura.

Virtually all science articles move from literature review through the development of testable hypotheses grounding the description of research design to what is called the specification of models—possible causal sequences to be "estimated," evaluated with data, in the next section of the paper. I have actually skipped over a stage—hypothesizing—present implicitly in almost all articles but explicitly only in some. The logic of science reconstructed by science itself is supposed to unfold this way:

Theory→Hypotheses→Design→Specification→
Estimation→Discussion

Whereas every paper has a literature review, design, model specification and estimation, and a closing discussion, only some explicitly break down the literature review into testable propositions or hypotheses. This is most frequent when the articles are not databased but only suggestive; these are sometimes considered "theory" articles, as we see in chapter 9. Many empirical articles short-circuit the formulation of explicit hypotheses simply because everyone knows that literature review segues into the discussion of research design followed by model specification and estimation—the heart of the data analysis. Yet, if not an explicit text, hypotheses are always a subtext. Occasionally, they are made explicit, as below:

> This review of the research inspires several hypotheses.[21]
>
> In order to test models, they need to be translated into, or serve as inspiration for, the development of testable hypotheses.[22]
>
> It would be impossible in the present exercise to develop hypotheses for all of the causal paths delineated in Figure 6. As an alternative, some of the more important motivational dynamics can be outlined, with an eye to developing a number of plausible hypotheses.[23]
>
> It would be possible to generate additional hypotheses, or to decompose the ones offered into simpler propositional statements, but the discussion above can at least illustrate the theoretical strategy that I propose.[24]

The second passage summarizes well. Either science needs explicit hypotheses enroute to testable models or it needs hypothetical "inspiration" for the estimation of models – symbolic constructions of the suspected relationship among dependent and independent variables. Dependent variables are caused by independent variables; for example, in a study of the impact of race on economic inequality, the dependent variable might be annual income and the independent variable might be racial and ethnic categories. Sometimes writers actually state their hypotheses in propositional form:

> Hypothesis 1: Anti-competitive corporate behavior will result from a squeeze on corporate profitability which may emanate from the firm, its respective productive market, or business cycle downturns.[25]

If actual hypotheses are not stated, operational definitions help turn the paper's topic into researchable terms.

> Our operational definition of utopian literature comes from Sargent (1979, p. xvii), who defines it as prose fiction that contains "a fairly detailed description of a social system that is nonexistent but located in time and space."[26]
>
> In operational terms, these arguments suggest that there should exist an empirically verifiable relationship between indicators of dependency/world- system position and the sort of intranational factors that are seen to be the proximate determinants of urban growth.[27]

All these preparatory constructions link literature review with the fine-grained analytic work itself. Whereas most papers lurch directly from literature discussion to the description of research design (followed by model specification and estimation), some pause to hypothesize. Hypotheses are prose formulations of the estimated models, constructions of how variables fit together in one-way or two-way relationships. As prose constructions, hypotheses allow the writer to remain a little longer in prose-land; most scientists are too impatient to linger long there. They want to jump right into the analytic work – research design, model specification, model estimation, concluding discussion.

In the literary construction of hypothesized causal models – model specification – the writer reconstructs the work he or she did in forcing the data into conceptual shape. Data as such tell nothing; they must be scripted into meaningful patterns. For workaday science, much of this work is done by mathematics; specified equations are hypothesized relationships among variables to be evaluated with reference to data. Of course, not every science article uses specified and estimated models. Some lag farther behind in mathematical sophistication than others. In every instance, though, the literary challenge is to reconstruct the logic of inquiry in showing how the collected data were made to speak. Rarely does the specification of models precede the data collection; most social scientists do not collect data at all but engage in secondary data analysis—reanalyzing the data of others. Few researchers can afford the time and money to do massive original surveys. Instead, they borrow or purchase datasets (large amounts of data entered on computer tapes) that they then plunder for further meaning. Thus, model specification reconstructs the way the writer thought meaningfully about the possible patterns of the data themselves.

> The first empirical task was to specify a regression equation which accounted for multi-level sources of profit-squeeze—the firm, industry, and business cycle. Given the longitudinal design of this study, variable attrition was a common problem.[28]

"Longitudinal" is a word meaning over time. The methodological discussions of model specification are no less technically tedious than the discussion of research design; if anything, they are more so. The problem is not that science does not use common-sense terms for things; after all, the things science is talking about are hardly commonplace. Jargon is not the issue; technical terms like "longitudinal," "variable attrition," and "regression equation" are code words understood by insiders. What is worth noticing is the extent to which model specification continues the extremely close discussions in the description of the research design of an analysis. The prose is written in code for insiders; what matters most is technical nuance, not broad-gauged issues of theory.

Even a cursory glance at the methodology sections of papers will indicate contemporary science's deep investment in technical issues. Increasingly method's topic is method itself. While findings matter, the sections of research design and causal modeling are often the bulkiest of the paper; that they are written virtually in hieroglyphics only compounds the problem. Readers, to plow their way through the methodology sections, must take time. Whereas they can glance through the literature review and successfully catch the general drift, method does not disclose itself to quick readings. This is part of science's aura: It consumes attention (and, as such, diverts attention from other issues). American sociology models itself on the natural sciences but fails to understand that, for natural science, methodology is not the most important text. As an infant science, sociology consumes itself with technical issues. Only in this way can the discipline gain equal stature to the hard and applied sciences, a crucial issue in a time of retrenching universities.

Models are presented complexly; their presentation is couched in technical argot, symbolization, and methodological circumspection. Doubts are anticipated and answered. Whereas the initial theory section of science papers rides roughshod over intellectual nuance by sausaging sources together in authority chains, science takes its time with the real issues of model specification. Nothing is more important in science today.

> One of the problems with adjusting equations for heteroscedasticity is that the correction procedures may not offer equations that are significantly better than the OLS originals. This is true if the disturbance in the error term is more complex than that which is modeled, or if more than one variable is correlated with the residuals. Moreover, heteroscedasticity in OLS equations may be related to misspecification (i.e., the equations are missing a crucial independent variable, are nonlinear in form, or there are measurement errors which are cumulative over time, Koutsoyiannis 1977). Substantively it can be argued that the degree of heteroscedasticity in these OLS estimates is a minor problem.[29]

Admittedly, this is from a footnote. But consider this:

> First, it is widely known that in pooled regression models, OLS techniques yield inconsistent estimates because they fail to correct for autocorrelated disturbances...A second estimation problem arises from heteroskedastic residuals caused by the enormous size differences among the states.[30]

"Heteroscedastic"? "Autocorrelated"? Just because I find these words unfathomable but understand reification, mediatization, surplus value, and negation of negation only indicates my limited intellectual formation. Jargon is an issue, but it is not the main issue. At a time when ordinary discourse has been incredibly degraded, it is hard to avoid expert languages, even if one wanted to do so. The problem is that analytic substance is traded for these elaborate concerns about method. How one studies something and then evaluates the results matters more than what one finds and what it means. Where ordinary language degenerates into media-purveyed cliche,[31] technical language goes out of control. It loses the ability to think its relation to the topic at hand—here, the topic of science articles. Unless the topic expressly is methodology, these obsessions with method (here, model specification) are only that—obsessive.

Of course, it is easy to say this from the outside. I do not use such words as *regression* in my work, except to mean "backsliding." Fine. We intellectual elites are allowed to speak in our native tongues. On a scale of obscurantism, Hegelian Marxism does not fare better than inferential statistics and regression analysis. Indeed, one might observe that we all regress, or at least miss the point. Yet there is a difference. Hegelian Marxism tries not to lose sight of the totality—the whole of world history resulting in the bad present and diminishing the possibilities of a different future.[32] Hegel was the first philosopher to address totality; Marx merely historicized him. I am not proposing a better intellectual slogan—Hegelian Marxism for scientific sociology—but observing that obscurantism is not the overriding issue here. Rather, science as literary presentation has gotten bogged down in the logic of methodology and mathematics. As a result, substance suffers.

Responsible methodologists would insist that method is substance—that only dilettantes analyze data without understanding technique thoroughly. I grant the point. But positivists believe that there is an inherent logic of scientific methodology requiring these sorts of picayune discussions.[33] Science writes as if the final epistemological arbiters are the rules of sampling, statistical inference, mathematical modeling, and matrix algebra. But these are not logics imposing themselves, willy-nilly, on science's narrative work. These are only tools, aids, semiotic media conveying the ways data can be shaped to suit an argument. Math cannot decide arguments any more than citation can. The issues must be argued, prosefully. Method only drives the discussion of data when its rules of evidence, inference, and modeling overtake our abundant interpretive reason. Method does not make the data speak; we do.

Modeling: Semiotizing Hypothetical Reasoning

There are two ways to view the figural work littering science articles. Either these nondiscursive gestures, including formal models (equations), number, and graphic designs, are believed to proceed under certain nonarbitrary rules binding on all users[34] or they are viewed merely as illustrative.[35] Journal science assumes the former; I argue for the latter. This does not mean that one can credibly semiotize one's argument in any fashion; while language's limits are not absolutely fixed, neither are they infinitely broad. Science, particularly mathematics, is a form of rhetoric—argument, analysis, answer. Many would disagree. I suspect they disagree because they view mathematics as a principled system and not as a mode of reasoning and rhetoric. They want to view mathematical science as hard-and- fast because the universe otherwise threatens to decenter. In times of ideological, even metaphysical, doubt—political "isms," the threat of nuclear war, transmuting values, AIDS—nothing comforts like system. Science promises respite from the fuzziness of the times, even from the indeterminacy of its opening literature review (itself already highly constrained by discipline).

But while mathematical reasoning has certain principles of consistency and logical exclusion, it is important in written science as a mode of argument—rhetoric. Dogmatic scientists cannot see this because they refuse to; otherwise, the indeterminacy of the world comes crashing in on them. Apart from the existential security afforded by mathematized science, math in itself creates a science aura, a crucial literary strategy of ideology in these times. A lawful world can be reduced to formal models, elegantly circumscribing human freedom within deterministic models of the universe. Math functions politically as a worldview ensuring determination—loss of autonomy. Rhetorically entered into the text of science, mathematics reflects and reproduces a freedomless universe; this is where its metaphysic of determination becomes crucially political.

Thus every mathematization and semiotization of experience constrains it still further. Working scientists do not recognize this about their literary work; they would probably deny it. They believe that mathematics has certain incontrovertible principles deciding analytical issues in their own terms. But this makes two mistakes, the one historical and the other literary. History resists all efforts to reduce it algorithmically to deductive principles; tomorrow awaits, and thus thwarts deterministic logic. Lawful social science, in imitating physics, commits a metaphysical mistake; although social being is matter, it is also composed of intellect. We are not fated to repeat past mistakes, forgettings, oppressions.

The second mistake involves interpretation.

> We used regression in order to evaluate general trends in differences in psychological well-being between blacks and whites over the years.[36]
>
> By using recursive equations, we assert that time together is influenced by work schedule, life cycle, and sociocultural factors. This reflects our theoretical position that work schedules affect time together rather than vice versa (or some more complicated reciprocal pattern of influence).[37]
>
> The finding of a statistically significant positive relationship would support the argument that level of dependency influences level of sectoral inequality or that, ultimately, the effects of international forces on urban growth are mediated by level of sectoral inequality.[38]

Methodology cannot decide interpretive disputes; it only suggests rhetorical strategies with which to formulate interpretations that seem credible to other language users. As metaphysics, methodology decides nothing; the data resist their reduction to ontological ciphers. As semiology, methodology can do much literary work, suggesting convincing insights and interpretations. The mechanical operations of statistics and causal modeling do not produce airtight interpretations but only facilitate the proseful reconstruction of interpretations in an economic way, substituting an elegant semiology (figure and number) for words.

But the words are always outer limits for number. Science articles begin and end in prose. They must because methodology cannot decide intellectual arguments. The problem is that science writers forget this; instead, having become so accustomed to methodology as a norm, they take methodology to be the essence of the logic of discovery. Mathematics is not a substitute for thought but only one of its semiotic agents. Where I would say that science semiotizes hypothetical reasoning figurally, science might say:

> A simultaneous test is possible using a panel regression model that has been employed in other quantitative, cross-national studies (see Bornschier et al. 1978; Rubinson and Holtzman 1981). This OLS model allows all potentially relevant variables to be tested against each other. The model regresses each dependent variable at a later point in time (1980) on its value at an earlier point in time (1960) and on the independent variables. A panel design (1) tests the stability of the dependent variable over a specified time period (1960 to 1980 in this case), and (2) controls for the prior effect of the dependent variable and thus prevents false inferences due to reciprocal causation.[39]

There is no correct way to write or figure reason, only a number of available rhetorical strategies. I talk my way, and journal science talks its way. We probably try to persuade different people. Our aims are at once similar and different. We both seek consensus; yet science would reproduce a selfsame world by modeling it lawfully, while I would transform the world by reading science itself as a powerful political language concealing its authorial desire to bring about a certain state of affairs.

Science resists this reading, but unsuccessfully. It hopes to capture the entire terrain of discourse—in this instance whole academic disciplines in which people talk like this—and thus exclude other accounts. In disciplining philosophical and poetic talk, science would vanquish its opponents and then carry on with business as usual. Few radical theorists read mainstream science;[40] we dislike it, distrust it, despise it. But in letting science talk as if it were really an arbiter of intellectual differences, we fail to uncover it as the fiction it is. Too frequently we forget that our own "science," whether historical materialism or feminism, is fictive too. But reasoning cannot escape its bondage to rhetoric. Indeed, I celebrate the narrativeness of reason as a fortuitous vehicle of community building. In writing writing as the social text it is—even, no especially, science—we open it and its metaphysical mistakes to a lay community of speakers. We can all do science; we must if we are to confront the fact that the monopoly of wealth is paralleled and propped by the monopoly of discourse. A peculiar mode of intelligence in its own right, science does not exhaust intelligence. It must be made to share it.

Once we read science as the literary account it is, semiotizing reason not to decide philosophical debates from an Archimedean vantage of unchallenged privilege, we make it available to many users. The left is not stupid; we can learn the techniques of math and science. Indeed, we must if we are to understand the complexly mathematized, mediated, and overdetermined world order of the present. How otherwise can we understand international economics, the rate of oppression, science and technology as ideology? Quantitative method is not immoral, both capitalist and sexist; it is only a way of talking about the world. It is no more or less privileged than theoretical talk; indeed, mathematical elegance has a lot in common with dialectical systemizing. To cite an inspiring left-wing authority, where would Marx have been in *Capital* without recourse to semiotica of economics and mathematics?

Yet it is impossible for radicals simply to use mathematical methodology without being used themselves. Methodology is not only a literary expression; it is also a political act frequently attempting to close the universe of historical possibilities. Radicals cannot simply write science in the conventional format of positivist journals, for that subtly provokes a world in which we live our lives as they are seamlessly modeled for us. Methodology inspires thought only when it is unhinged from the driving positivist text inscribing the world in itself as pure presence. Figure is not representation but only literary construction. As such, the whole apparatus of journal science including literature review and an ontologized multivariate analysis conspires to make methodology a medium of thoughtlessness. Method is not value free; but as literature it can be employed illustratively and imaginatively. I return to this theme in my concluding chapter.

Modeling as Text

I have begun to talk about modeling; let me turn more closely to how it talks about itself. Science wants us to attend to the technical details; I want to observe the rhetorical cadences, constructions, slips. Model specification can be read as a rhetorical mode even if it occasionally affords insights into the world on its own terms. It may do so in spite of its pretension to be a vehicle of absolute knowledge. It is difficult to believe that the following incantations have become so central to journal science given their indirect relation to the world at hand. But that is their voodoo, their aura. The less we can puzzle out the representational implications of causal modeling, the more they seem to let the world reflect luminously on the page. If we do not understand the empirical world very well, then we can believe that its essence is captured in the figurings on science's page. Indeed, if we did understand the world adequately, we would not believe that the semiotics of figure did more than spike the imagination. The world is not ontologically numbered.

> This analysis uses a time-dependent ecological structure of the type commonly used to model population fluctuations among competing species within a biological ecosystem. This type of model has been described in the social scientific literature by...[41]

> Finally, we develop a more comprehensive causal model that incorporates both intellective and non-intellective aspects of personality, paralleling a model previously developed for adult male workers (Kohn and Schoenbach, 1983; see also Kohn and Schooler, 1982).[42]

> One possible model for pre-college schooling is the censored regression...model (e.g., Tobin, 1958; Maddala, 1983) which recognizes that observations are concentrated at 12 grades and that, whereas the effects of father's schooling may be linear given that a son has 12 or fewer grades, the total effect of father's schooling is inherently nonlinear.[43]

> The effect of father's class and education on son's and daughter's education is estimated by ordinary least square (OLS) regression. Father's class is coded as four dummy variables...[44]

> The variables are entered sequentially into a logistic regression (because of the dichotomous dependent variable)...The first criterion for causal order is timing of the event. The second criterion is theory. When timing and theory cannot resolve the order, we decide by fiat.[45]

> The dependent variable in the multivariate, logistic regression model is P's choice of O as Best Friend or non-Best Friend.[46]

> Since a two-indicator measurement model is underidentified, we embed this simple measurement model in a larger non-causal model that contains all the non-psychological variables that will enter into the causal models that are described below.[47]

Modeling appears to proceed according to an inherent logic reflecting the logic of the social world. This is false. Models are fictions, speculated by the author and then evaluated with data in the paper's next section (see chapter 6). Of course, the text of model specification suggests otherwise; even the word *specification* connotes a thoroughly technical process. Models are merely hunches about the patterns of the social world. Informed by insight and experience, they organize research, allowing data to be compartmentalized in intelligible ways. Methodology pretends that it reflects deterministic processes out there. The world is thought to govern research strategies, here modeling. Specified models are simply guesses about causal relationships among variables. Even that language presupposes too much technical familiarity. Variables are things whose values can change—temperature, pulse rate, income, years of education.

Models semiotically summarize guesses in equations. Dependent variables are linked with independent variables in a way that allows them to be tested against data. The strengths of relationships among variables are expressed in correlations simply describing the likelihood that change in one variable occurs at the same time that another variable changes. Correlations only suggest the likelihood that variables vary together; only so-called perfect correlations suggest absolute causality: When X changes, so will Y—invariably. Models are equations suggesting correlational, thus implicitly causal, patterns among variables. A specified model might suggest that income is a function of level of education plus gender.

$$\text{Income} = f(\text{Education} + \text{Gender})$$

Data are then used to check the strengths of the relationships. That occurs in the paper's descriptive section, where the model is estimated, checked against data.

One does not specify just any model but only models corresponding with one's sense of the data, the existing literature and the world as one understands it. The first quote actually names its model—"a time-dependent ecological structure"; or, in another instance, the authors' model parallels "a model previously developed for adult male workers." The creation of models has reached a certain technical level of development; one thus couches one's talk about them in rarefied terms: "two-indicator measurement model," "multivariate, logistic regression model," "dummy variables," "logistic regression." Multivariate analysis—analysis addressing interrelations among variables—is incredibly complex when model specification and model estimation are viewed almost entirely in technical, mathematical terms. Now there is nothing inherently wrong with this kind of work, any more than there is anything the matter with pure mathematics. But science writers do not regard their work as literary; they do not admit that models are fictions conjured by the writer to make tentative sense of complicated, messy data themselves imperfectly reflecting the social world.

Instead, science writers write as if these methodological problems have self-evident technical solutions. If a model is "underidentified," it can be fixed mathematically or technically.

> Log-linear analysis was developed as a more powerful technique for the analysis of multivariate tables of categorical variables. It was specifically designed to deal with some of the problems that plagued users of crosstabulations. One of the major drawbacks to the crosstabulation approach is the difficulty of controlling for the effects of more than a single other variable. This is often due to small sample sizes within cells of a crosstabulation entailing further control variables.[48]

Science writers believe that methodological technique in itself solves research problems. "Log-linear analysis" is "a more powerful technique for the analysis of multivariate tables."

> An alternative to conventional panel regression entails the use of instrumental variables. Markus (1979, pp. 51-54) outlines a procedure whereby the lagged dependent variable is regressed on the other independent variables and the new value of this variable (the "instrument") is substituted for the original lagged dependent variable.[49]

I do not doubt or deny that mathematical and methodological problems can be addressed in their own terms. But the work of model specification—semiotizing one's research hunches in the elegant syntax of multivariate regression analysis (i.e., assessing the relative impacts of independent variables on a single dependent variable)—unfolds totally within technical logic. Science discusses "procedures," not substantive findings. The methods sections of science articles are by far the longest, brimming with technical suggestions and caveats. But there is no absolute answer to an interpretive problem, merely a number of different expedients.

One would think that science articles are about the method and mathematics of model specification, measurement, estimation; at least the amount of attention given to methodological issues, especially model specification, seems inordinate given the brevity of science articles in the first place. Admittedly, people on the inside want to know about how studies were done and then interpreted. But hypothetical reasoning and data interpretation have been so semiotized—so mathematized—that the mathematical logic of technical elegance has virtually displaced the search for an objective truth. The technician would say it has to be this way lest sloppiness erode the advancement of science. Yet there is no clear technical progression from one article to the next; no matter how sophisticated in quantitative analysis, different people do the same work differently—different topic formulation, different data, different models, different interpretations. Methodology has become a babel of tongues.

I found the following passage in what could have been a very interesting article on secularization and church attendance.

> Log-linear models are well suited to the separation of substantive change from sampling error. We fit standard hierarchical models (e.g., Goodman 1970, 1972; Fienberg 1980) to test the null hypothesis that year per has no effect net of the changes in papal authority and sexual othordoxy. Our search is guided by the *bic* statistic introduced by Raftery (1986a, 1986b). Extensive model search through large numbers of cases frequently leads to overfitting, that is, the inclusion of terms in the model for which we have little or no substantive interpretation (except a post hoc invention). In applying *bic* one chooses the model with the largest negative *bic*. The search for a preferred model of the complete 5 x 5 x 5 x 2 (attendance by sexual orthodoxy by papal authority per year) table is summarized in Table 8.[50]

The methodological pyrotechnics are coupled with the degeneration of prose; exactly what are we to make of "to test the null hypothesis that year per has no effect net of the changes in"? Since when did "year per" become a noun and "net of" a prepositional phrase? And do specialists really comprehend what this means?—"In applying *bic* one chooses the model with the largest negative *bic*." In any case, methodological sophistication substitutes for real insight; science is driven by its compulsion to become mathematics.

Occasionally, the truth slips through; science admits the literary nature of "model specification" or simply hypothesis construction, albeit in a language purposely made to read like science.

> The composite model developed here is only an intermediate step. It is possible to construct several different composite models and then assess which is most parsimonious and

> isomorphic with actual empirical processes. Figure 6 represents only a preliminary theoretical judgment as to the critical dynamics behind motivational processes, but it has the virtue of being explicit about the crucial variables and their causal effects on each other.[51]

Of course, this is a "theory" article, the topic of chapter 9. The writer apes the technical prose of science when he says: "It is possible to construct several different composite models and then assess which is most parsimonious and isomorphic with actual empirical processes." Isomorphic with actual empirical processes? He means: Descriptively adequate (or some such). He does not say this not because he cannot write English well but because methodology has become a mind of its own, a way of reducing the logic of scientific discovery[52] to technical steps.

The more science reads like what it conventionally takes to be science, the more it can defend its own results. Thus the esoterica of research design and model specification squeeze out comprehensible thought. If methodology is simply a way of reconstructing one's intellectual labors in coming to and then working through a problem, then a certain measure of semiotization is perhaps acceptable, at least for illustrative purposes. And if there was real continuity in journal science, writers following up on others' methods and findings, then one could perhaps hope to advance disciplines through technical innovation. But methodology is more today; it is everything, especially when figure displaces prose as a narrative vehicle. That is what is happening in journal science; technique is fetishized and semiotized precisely because that way one can reduce authorial presence to a bare minimum—at most to the choices authors made in their solutions to certain trivial technical problems.

The research design/model specification section of the paper prepares us later for the utter eclipse of text with figure and number. Model specification puts thought into equations; its aim is bivariate and multivariate regression analyses—assessing the impacts of variables on one another. As the passage above admits, thought can go anywhere; equations encoding causal models proliferate. Data put the brakes to decentered thought without necessarily creating a new center. The world is not mathematically lawful, although we can develop mathematical laws merely as poetry. In effect, methodology as a literary effort prepares us for the figural and interpretive work yet to come in science articles. Methodology contributes itself to the science aura—the notion that the text in question validly reflects a lawful world outside itself. This is fundamentally untrue inasmuch as positivist methodology fails to account for its own contribution to bringing about that world. Without it, the world remains indeterminate; people fail to play the roles scripted for them by science. Science interposes itself between history and the world in order to enforce the patterns suggested in its equations. At root, science describes a deterministic universe, knowing secretly it does not exist. It knows its equations to be false because people can choose against them, failing to reproduce the patternings of class, race, and gender constraining them for the past two thousand years.

This is not to say that people simply choose their own oppression, unnecessarily. Oppression is still oppression; its nature is its imperviousness to the will of the victim. But victims also choose it when they fail to understand what is happening to them as historical; instead, they view their own deprivation as a fate captured (and thus redoubled) in the solemnities of social science. Some even come to love it, or at least to relish its representation in the artwork of causality; we call them positivists. In secretly knowing that they do political work, they fail to see that they are victims too. It is imaginable that free people would do science and mathematics as vital self-expression. It is unimaginable that they would sacrifice their own authorial subjectivity to the literary apparatus of science and math in the meantime. Of course, we call this professionalization. Professionals, circularly, are those who can read the writing of other professionals. Semiotic ability passes for intelligence in the scientized, academicized present.

My department hires "technically competent" faculty members who can read and write the science journals. We value people who can publish in the "major" journals as well as obtain outside research grants with which to support the academic profit center. These are only literary abilities. I recently read a thin monograph on "applied regression" analysis; it is possible for this sort of work to be understood just as it is possible to read and write in a foreign language. Eventually the dense text yields its meaning more quickly; we recognize colloquialisms, twists, mistakes, habits and can read through them. The same applies to journal science. After awhile the technical lingo makes sense—not necessarily as precise methodological recommendations or descriptions but as the sense of what people do who do science.

Of course, it is tempting to anthropologize science just as we cherish the natives whose languages and customs we learn. Ethnomethodology is a methodological and theoretical variant within sociology that exhibits the ordinary sense-making practices imbedded in all language systems.[53] Even when people make communicative mistakes, they succeed in having themselves understood. I read science in much the same way; it is a routine practice inhabiting a certain narrative text. Yet, in itself, that is only half the story. We could say that everything is language, everything text; our deconstructive work simply appreciates other versions of reason. But some languages, certain cultures, oppress as well. Science is literary work that, in peculiarly concealing its authored nature, deceives and thus oppresses. In modeling a causal world, science would cause it. The problem is that the world is open; irreducible to mathematical laws, it could be different and thus could be understood differently.

Some would make it epistemologically illegal to do math and science in the realm of the social. I am not among them. I would proscribe only the literary strategy called positivism—most journal science. Number can usefully provoke imagination and thus freedom; in themselves, ordinality and cardinality are just rhetorics summarizing certain events or ways of being. But method becomes madness when it attempts to decide metaphysical issues such as causality. Regression models are fine as long as we realize that they do no more than tell a story, suggesting a version of the world to which they make a decidedly political contribution. Regression models cause intelligence to regress when they suggest themselves as an authoritative representation of an outside world. If data themselves are constructions—findings—then the semiotics encasing them are even flimsier. Mathematization does not drive out caprice or error; to think otherwise is a mythological stance unbefitting to science.

So it depends on what we mean by science. That people view their literary work as model specification is a fact; that they could different work under the name of science is also true. Anthropology is partly right; in itself, custom is fascinating. But sometimes custom causes undue misery—foot binding, racial segregation, positivist mind. Then it must be understood for the contingency it is, and opposed. At least, we must read custom—here, mathematized science—as evidence of the openness of the world. Scientists are telling us something by doing the work they do. They are saying: Avoid politics, poetry, passion; these are false idols. Let things be. In understanding how science understands understanding, we understand better than it. Better, we improve science. No matter what—no matter how many lines it devotes to technical issues—science cannot drown out its author struggling to be free of these conventions. Or to be free, period.

Chapter 6 Science Aura II: Findings Found

Findings as Literary Productions

The elaborate discussions of research design and model specification—methodology—are followed by an equally elaborate section in which the paper's findings are presented. No less technically detailed than the preceding methodological sections, this section engages in what is called model estimation, putting numbers into the postulated equations suggested in the preceding section. This section of the science paper is descriptive, evaluating the model specified in the earlier section in the light of the paper's research findings. Yet the data presented in this section are not really discussed or evaluated in a larger sense; that must await the final, brief, section of science papers typically called the "discussion," often no more than a few paragraphs, maybe a page.

One might expect that the model-estimation section would simply duplicate the earlier model-specification section with the addition of data, and in some senses it does. The degree of technical specificity is no less; larger conceptual and interpretive issues are avoided. If there is controversy, it is about technical, methodological meanings. Yet the findings section differs from the methodology section in a crucial respect. The text is suddenly littered with figure—number, graph, chart, table, arrow, scatterplots, virtually everything but prose. I exaggerate here; the findings section does talk too. Yet its talk is larded with much more figural display than in earlier sections of the paper. This is the second part of the science aura; data are figured in a way that would represent the world's figure, directly reflecting the world beyond the mediating constructions of prose. While the use of number (a special case of figure) begins in the earlier methodology section— the specified model, for instance, expressed semiotically in mathematical code—the text is much less semiotized than it is in the findings section. Whereas the specified equation obtruded from the methodology section as a surprising gesture, initiating the science aura, in the findings section figural gestures are rampant. The argument is almost entirely semiotized, figured, enhancing the science aura. In this chapter I want to examine figure and number as important texts in their own right. First, let us talk of the text on its own terms.

> Model 8 presents what might be the most interesting result of this analysis. This model shows a significant interaction between probation and prison populations. Probation here shows a significantly negative main effect on jail expansion, as was hypothesized early on. More important, the interaction term shows that, after the enactment of probation legislation, the number of prison inmates exerted an upward pressure on jail populations (when PROB equals one, the prison coefficient becomes .62, and the parameter is significant).[1]
>
> The results of the goodness-of-fit tests are shown for the seven conditions in Table 5.[2]
>
> Being in a mixed work setting had a *positive* effect on the satisfaction levels of unmarried men ($b = 16.136$, $p < .01$, two-tailed test) and men whose wives were not employed ($b = 16.402$, $p < .01$, two-tailed test).[3]

At least prose subsists in these passages. A little detective work plus some basic training in methodology will disclose the meanings. Two of the three passages refer to figural work elsewhere in the text (usually near the quoted work, as is the convention in journal science). It is also notable that one cannot make it through any of the three passages without stumbling on either number itself or on technical usage ("goodness-of-fit tests"). The science aura endures in the findings section.

> Data in Table 3 reproduce a baseline human capital model of earnings with the addition of the structural change and gender composition measures. Model (1) regresses annual earnings (see Hodson 1985) for each sex on a vector of human capital characteristics, respondents' race or Hispanic origin, and a vector of control variables, including a selection bias correction factor λ.[4]

This jargoned passage indicates the basic work done in the findings or model-estimation section. The data (summarized here in Table 3) are run on a computer in analyzing the causal contributions of certain personal and social characteristics to annual earnings. The basic work done in the findings section is simply to insert data into the equations modeled in the methodology section in order to check them. Where specified equations are semiotized hypotheses, estimated equations are findings figured in number, table and graph for visual, hence analytical, impact.

> Equation (3) includes an age-proneness shift variable, which is not statistically significant. This variable is the product

> of a dummy variable, equal to one for the years 1966–84 and zero for 1947–65, times the natural logarithm of the age structure index.[5]
>
> The problem with a latent variable approach to an unanticipated association like this one between attendance and partisanship comes when we attempt to say anything authoritative about the substance of the latent variable. We have not observed the exogenous influence that, in the calculations that follow, turns out to be important for understanding the plateau in Catholic church attendance in the mid–1970s.[6]
>
> The murder rate series also exhibits a first-order negative autoregressive error process that produces an oscillatory pattern of trends predicted on the basis of these structural relationships.[7]

The findings are presented almost entirely in terms of the technical parameters established in the earlier methodology section. The results are analyzed in terms of the methodological requirements of model evaluation, as well as in the light of rules of statistical inference.[8] This is very important for the way it delimits the discussion of evidence. One must never make interpretations beyond the ken of one's data; one must adhere to the specified model and the rules of inference permitting one to make certain hesitant generalizations about the meaning of the data. In any event, journal science requires that one derive researchable hypotheses (semiotized in equations in the methodology section) from a literature review and then test those hypotheses in the findings or model-specification section of the paper. These are the ground rules; violating them breaks the norms of journal science.

The descriptive work done in the findings section is carefully crafted to relate directly to the semiotization of hypotheses in the preceding section on methodology. This would seem obvious; findings are to relate to what was sought in the first place. Yet this issue looms larger in chapter 8 where I discuss the way that journal science concludes. Here, already in the findings section, there is a hesitancy about drawing conclusions; indeed, the findings narrowly confirm or disconfirm the particular suspicions turned into equations in the preceding methodology section.

> The evidence for Blake's interpretation comes from a regression analysis of son's elementary and secondary schooling, coded as actual years completed if he attained 12... years. The ordinary least squares (OLS) estimates of the father's schooling effect in the second panel of Table 1 increase with sibship size much more sharply than for total grades of schooling.[9]

> Therefore, there is some evidence to suggest that the effects of international position on urban growth may well be mediated by level of sectoral inequality.[10]

Usually the findings section simply reports and figures findings, nothing more. Where it hints at causality, it is extremely hesitant about drawing conclusions unwarranted by the particular data. Manuscript reviewers and journal editors frequently chide authors for "going beyond the data"; one can see ample evidence of equivocation when writers couch their findings. And the nature of this equivocation, as we can see, is basically methodological in nature. "The evidence for Blake's interpretation comes from a regression analysis...," not from thought. Technique does the work of interpretation—what little there is of it in journal science. Let me give a telling example of this:

> The nonsignificant effect of total foreign investment contradicts previous studies (Kentor 1981; Timberlake and Kentor 1983; Bradshaw 1985b) which show a positive relationship between foreign capital and urbanization. As demonstrated in this analysis, such an association does not obtain when other important variables are included in the equation.[11]

The key word is "nonsignificant." One might read the authors to be saying that they found little relationship between foreign investment and urbanization, in this contradicting "previous studies" cited. On the other hand, they might be saying that the relationship between foreign investment and urbanization is *statistically* insignificant. That is probably what they are suggesting. They go further to argue that the shortcomings of other research are not due to stupidity but simply to underspecified or misspecified equations—wrong causal hunches leaving out "important variables" from the equation. Their discussion of findings is driven by technique—the nature of their equation—and not by a more reasoned argument about what is really happening in the relationship between foreign investment and urbanization, one of the less trivial topics treated by journal science.

Intellectual differences are reduced to methodological quibbles. The result is that intellectual debate is discouraged in favor of technical innovation.

> Finally Ornstein (1982, 1984) and Palmer (1983) may have found that the number of interlocks connecting firms stimulates reconstitution, while we did not, because both authors did not measure the organizational and class interests that firms and

> their directors have in interlocking; in the case of the former, financial constraint, and in the case of the latter, both financial constraint and headquarter proximity.[12]

The problem with the other authors cited here is one of measurement. The analysis attempts to improve on them by making better measurements, using improved techniques. This is bound to happen when science articles spend so much time constructing their problems as technical in nature; the research design and model specification sections are almost entirely consumed with technical virtuosity, making it inevitable that the findings would be discussed in minutely methodological terms and figured beyond words themselves.

The construction of research problems virtually decides their subsequent intellectual treatment. The narrower and more methodologically rarefied the topic, the narrow and more rarefied will be the subsequent report and discussion of research findings. I suspect that many insiders spend most time poring over the research design and model-specification sections, ferreting out technical nuances. They can scan the findings sections quickly, especially because the findings are nearly always figured in readable tables, charts, and graphs. Although the findings are important in an obvious sense, they are less important than method when the paper's topic is incredibly trivial to begin with. Given narrow trivialization dictated by the ironic nature of mainstream science, the only life left in science for insiders is technique.

As a member of a mainstream social science department, I listen to frequent discussions about job candidates and their presented research. Rarely do these discussions concern the actual findings; instead, they are preoccupied with methodology. Debates flare about the esoterica of these methods, resembling the level and tone of discussions in journal science. A colleague said that he opposed a job candidate because his data were not "clean." People are either methodologically sophisticated or not—the more the better. In this sense, journal science educates professional judgment, and vice versa. After all, journals are run by and for academics. At some level, journal science reflects (as well as reproduces) what is going on in the professoriate.[13] Academics promote the science aura and are themselves blinded by it.

We often assume that prose and figure are fundamentally different rhetorics, just as we take for granted the epistemological distinction between the social and natural sciences.[14] Thus, as I have done so far in this book, we read prose for traditional prose meaning and leave the discussion of figure for another section. While I too follow that logic, I am uncomfortable with it. The methodologization of language turns words into figures themselves—lifeless signifiers disinviting their dialogical interrogation. We humanists assume that figure and number themselves degrade meaning and thus, indirectly, lead to civilizational decline. While true, there is more to the story. The semiotization of discourse affects prose itself, not only prose displaced by their semiotic equivalents (e.g., equations for hypotheses). Prose itself becomes figure when it only talks technically, searching for solutions to its self-contained counterpoint—dialectic—in methodological rituals.

Of course, just as discourse is increasingly methodological in its nature, so methodology is also discourse. There are no purely or inherently technical solutions; statistics, regression analyses, research design cannot solve problems in themselves. They appear only to still debate and diminish dissension in an age of science. They narrate the argument, nothing more. Yet the methodologization of prose makes it more difficult for old-fashioned humanists simply to hate number; even prose that does not rely directly on number numbs intelligence when it sinks all its intellectual capital into what appear to be merely technical solutions to problems. Prose itself becomes technique under the science aura. Words are not crafted but simply juxtaposed serially in whatever strings produce the best science effect, connoting expertise and methodological professionalism.

> From a methodological standpoint, the potential for such differential effects raises an intriguing possibility: weak or null effects of structural characteristics on *overall* rates of criminal victimization could conceivably conceal appreciable but counterbalancing effects for different types of victimizations.[15]

> Table 3 displays the estimates of the ordered logistic regressions for the probability of "searching to quit." The first panel gives the CV model, with the authority-cost coefficients constrained to zero; the likelihood-ratio statistic is highly significant ($\chi^2 = 288 > 38 = \chi^2.001$). The middle panel gives the authority-cost coefficients with the capital-value coefficients constrained to zero.[16]

In the findings section of journal science, prose loses its mind. It becomes a figural gesture in its own right, connecting technical terms and number. In this sense, where I argue that figure becomes a telling text in its own right, so text becomes figure, purging itself of dialectical snap and bite—the authorial sense that it could have been written differently. When I posed this deconstructive reading to a friend who does quantitative research, she explained to me that "no one cares" about literary stylization; I was referring to an influential methodology treatise published by a major social science monograph publisher in which the authors used the term "plug in"

[data]. While no grammatical authoritarian, I proposed a reading of methodology and journal science as a witting and unwitting political subtext purging the analysis of prose able to interrogate itself and thus invite other versions.

This is another way to read jargon.[17] While obviously code indicating professional standing, science is a special jargon in that it would replace itself with pure figure and number; words gesture whereas science makes believe that the point of writing is only to communicate a body of findings. But in figuring the findings—even in technical prose—science makes deep assumptions about the representational nature of truth. It is assumed that the world can be better drawn than spoken, suggesting an approach to objectivity heavily reliant on representational identities between words and things. Science's jargon identifies itself virtually with the world to which it pretends to bear only a nonconstitutional relationship. This is false when science also helps constitute that world reduced to technical and figural gestures. In any event, a reading of the findings section of science articles suggests that the traditional humanist critique of mathematics is *under*extended; prose itself becomes a numbing piece of technique, thus losing the ability to reason.[18]

The Text of Figure

It is impossible to understand the aura of method, figure, and number without addressing the role of nondiscursive expression in science: tables equations, charts, graphs, scatterplots, symbols—the whole semiotic force field of science. While the findings section combines prose and figure, figure predominates; indeed, as I have just shown, the prose is figure itself. Science moves further away from the proseful literature reviews with which it begins. The findings section sees the eclipse of prose—better, its own figuring. The model of this figuring is figure itself, which in the findings section appears abundantly on the page. Prose is squeezed by the overarching tables, charts, graphs, and equations threatening it with extinction. And prose itself becomes figure where it resembles the technically thick constructions I have been quoting.

Figure rhetorically uses its shape to win arguments, to persuade. Science understands this perfectly well.

> The visual association between temporal trends in the age structure and crime rate series in Figures 3 and 4 is striking, particularly with reference to the substantial rise in the latter through the 1960s and 1970s.[19]
>
> Log-linear models were calculated to test the statistical significance of the patterns found in the graphs (see Table 2).

> They largely reconfirm the visual inspection of the graphs but suggest some caution in interpretation.[20]

"Striking...visual association[s]" win the day; "visual inspection" is confirmed by certain calculations. It is clear that representational clarity is text, and text only a buttress of it. Science figures its findings ostensibly to aid in analysis; graphic representation reduces the thickets of dense words. But the analysis supports the figural expressions, which are the primary texts. Text survives only as a remainder, an archaic residue. Science drives it out, or at least reduces it to technical gestures.

> The results above stem from parameterizing a square contingency table in a somewhat unconventional manner;...First, although we considered multiplicative models that mixed homogenous marginal effects with asymmetric association, we did not develop or discuss such models in detail. To the best of our knowledge, such models have not been previously considered.[21]

This passage is in fact less clear than many figures; overburdened with vernacular pretending to do real intellectual work, figure offers interpretive relief. At least, one can make sense of what is happening. Lines like those just quoted require the interpretive devotion of biblical scholars; methodology has become a monastic discipline open only to people religiously devoted to it. The ordinary disciplinary reader is as baffled as the layperson. Figures come as a relief; at least, one can exchange the illusion of comprehension for real understanding. That is how we read math when we come upon it in the bookstore. In leafing through it, we suppose that it might be understood; at least, it is indisputably rigorous, appearing to possess a disciplined relation to the represented world.

When prose does not disclose itself, we take offense. Words should not resist the attempt to understand them, translating them into other, more familiar words. Figure neither promises nor demands translation; it rests on the page as a version of the world itself—number, design, gesture making the world present, thus freezing it. The figuring of findings redoubles the science aura, convincing us of the indubitability of the account. In not having to be translated, interpreted, figure takes on a special textuality. Whereas words promise access to the world only through their translation into other words (themselves only existing in a mediate relation to things), figure delivers direct access—or so it seems.[22] We can understand this as the growing sophistication of social science increasingly adorning itself with the conventions of natural science. But that only explains circularly: Science is what science does. In fact, the semiotization of prose in journal science corresponds to, and thus deepens, the falling rate of intelligence generally. The mediate character of words frustrates in an age of instantaneity. Instead, we want immediate access to the things themselves, in the case of science the lawful object world. The more prose resembles figure, the more it figures in science's text.

Again, as with quantitative method, a particular case of generic figuring, the issue is not profigure or antifigure. One can gesture one's argument (as I have done briefly in chapter 2) without implying a presuppositionlessly representational relation between symbol and world; after all, figure can intend itself aesthetically, expressively. Figure is imperialist only when it drives out thought, suggesting a one-to-one correspondence between itself and the frozen world, thus freezing it further. I am not against science, only positivism. And figural gesture is a mainstay of positivism's literary strategy, enhancing science's aura in itself when figure and number are taken to be essential features of science and when figure as ontology intends a cleanly representational correspondence to the object world.

Figure circularly enhances the science aura when science is essentialized in its literary conventions; science is nondiscursive gesture, as such the most powerful text of all. Figure also affords a window on the world, freezing the apparently frozen world still further. Figure portrays because it is not thought to be freighted with assumptions, constructions, perspective. Of course, artists know this to be false; design designs, reveling in perspective as its authorial resource. Only science thinks figure and number disclose the universe unproblematically. Figure can be read narratively, as I am doing here. The irony is that the text of journal science is the least textless discourse today, aping figural techniques in order to suppress its authorial presence. Both word and figure are telling texts even if their relationship has changed; design no longer enhances prose but the other way around.

Science methodically conceals its own literariness to avoid its deep constitutionality. When it maintains that its prose is clean of metaphysics, it is second nature that figure exists only as a visual aid. Tables, graphs, and charts clarify dense prose. Any teacher knows this on one level to be true. The more the argument is semiotized, visualized, figured, the better it gets across, especially in a video age. My own students love in-class movies only partly because they dislike the dreary lecture routine. They also learn from them in a bookless culture.[23] Yet these visual aids become the real text, displacing prose or at least compelling prose to semiotize itself (thus requiring visual aids as a result).

The findings section is studded with references to figure.

> The results of estimating the models are found in Table 3, which contains parameter estimates corresponding to each period.[24]

> Figure 3 displays results that begin to address the problem of identifying partisan population migrations by relevant occupational categories.[25]

That is only the beginning. Not only does text engage with the figure surrounding it, it makes way for it. Figure starts early in the text, preparing the reader for the visual onslaught of the findings section. While proseful, the table shown in figure 6.1 suggests an inherent taxonomy to its analysis. It also translates "theory" into "variables," thus doing some of the essential methodological work of science. The table is titled "Summary of Variables Needed to Test Urbanization Arguments..." As such, the table promises more tables and figures to come—more *science;* it does not disappoint.

On the next page of the article, we get what the authors call an equation:

Level of Urbanization 1960—
[a + B(log GNP per Capita 1960)].[26]

This clearly raises the level of scientificity. Whereas before we were in the region of taxonomy and classification—early science—we are now in the land of mathematics. A similar equation appears on the page following this one in the article.

A page later, we hit gold. A table occupying at least the top 40 percent of the page summarizes all sorts of data on urbanization, population, and economic development (see figure 6.2). The table is devoid of words except for the far left column where nations are listed; the bulk of the table consists of numbers reporting data.

Apart from the obtrusion of number onto the page, it is also notable that the page contains little prose; the text is squeezed between the table reporting numerical data and a couple of footnotes at the bottom. Indeed, the text itself is interrupted by another equation. By now the scientific synchrony of the article is flowing. The page that follows this one in the article is a virtual duplicate, although now the prose itself is even more larded with number, jargon, and parenthetical citations, as such resembling figure itself. Prose is virtually nonexistent. Three similar pages follow. Finally, on the article's penultimate page, the dramatic obtrusion of figure is missing. Text is back in the saddle, or so it seems. This part of the paper begins the scanty "discussion" section, extending onto about a third of the final page. Where the article begins with the prose of literature review, it ends with the whimper of lame conclusions, hesitantly couched in the ambivalence of workaday science.

But things do not end there, at the final period of the discussion section. The last page of the discussion contains only a third of a page of prose.

Underneath the conclusion of the weak discussion, we find an "appendix" (see figure 6.3), figurally arranging the data sources for the discrete

228 AMERICAN SOCIOLOGICAL REVIEW

construct a variable that measures service-sector **labor** ***relative to*** level of economic development.

The variables of concern for each perspective **are listed** in Table 1. The calculation of each **variable** is discussed in the next section.

The three theories must be tested simultaneously to prevent spurious results. For example, an investigator cannot say that foreign investment in agriculture prompts townward migration without controlling for the pull of industrial employment and for the disparity in welfare between country and city. Likewise, an investigator cannot say that the welfare disparity causes townward migration without controlling for international economic factors (e.g., foreign investment and concentration of agricultural exports) and for industrial employment.

A simultaneous test is possible using a panel regression model that has been employed in other quantitative, cross-national studies (see Bornschier et al. 1978; Rubinson and Holtzman 1981). This OLS model allows all potentially relevant variables to be tested against each other. The model regresses each dependent variable at a later point in time (1980) on its value at an earlier point in time (1960) and on the independent variables. A panel design (1) tests the stability of the dependent variable over a specified time period (1960 to 1980 in this case), and (2) controls for the prior effect of the dependent variable and thus prevents false inferences due to reciprocal causation.

One objection to this form of panel regression is that the inclusion of a lagged dependent variable may preclude unbiased and consistent estimates in OLS (Markus 1979, pp. 49–54; Pindyck and Rubinfeld 1981, pp. 174–80). Specifically, the autocorrelation between the dependent variable and the lagged dependent variable violates the OLS assumption of zero covariance between the independent variables and the disturbance term. This will upwardly bias the effect of the lagged dependent variable if there is positive autocorrelation and downwardly bias its effect if there is negative autocorrelation. Despite this potential problem, most cross-national tests have continued to use a lagged dependent variable because it is a very conservative strategy. The lagged term usually is upwardly biased (because of positive autocorrelation) and therefore the effects of the other independent variables usually are reduced.

An alternative to conventional panel regression entails the use of instrumental variables. Markus (1979, pp. 51–54) outlines a procedure whereby the lagged dependent variable is regressed on the other independent variables and the new value of this variable (the "instrument") is substituted for the original lagged dependent variable. This two-stage least squares approach satisfies the OLS assumption of independence between the predictor variables and the error term, and it often is a sound strategy as long as the instrumental variable is highly correlated with the original variable (Hanushek and Jackson 1977, pp. 238–39). In a preliminary analysis for this study, Markus's approach was used to compare the results from OLS and two-stage least squares. The results were substantially the same in those equations where the instrument was highly correlated with the lagged dependent variable, suggesting that severe autocorrelation is not a problem in the analysis.

In order to establish the largest possible sample and thereby reduce the possibility of sample selection bias (see Berk 1983), this analysis includes all countries that meet a limited number of theoretical and technical requirements. First, countries must have complete data on all variables used in the analysis. Second, they must have had a GNP per capita of $1,000 or less in 1960, a standard used in other cross-national studies of underdevelopment (see, e.g., Delacroix and Ragin 1981). Third, countries must be considered nation-states and not city-states, a criterion that eliminates Hong Kong and Singapore. These stipulations are met by sixty-one underdeveloped countries.[3] Due to

Table 1. Summary of Variables Needed to Test Urbanization Arguments Relevant to Modernization, Urban Bias, and Economic Dependency

Theory	Variables Needed
Modernization	Industrial employment Economic expansion Relative urbanization
Urban bias	Disparity in welfare between country and city Economic expansion Relative urbanization Relative service/informal employment
Economic dependency	Total foreign investment Foreign investment in agriculture Foreign investment in manufacturing Concentration of agricultural exports Economic expansion Relative urbanization Relative service/informal employment

[3] The sixty-one countries include Algeria, Benin, Bolivia, Brazil, Burkina Faso, Cameroon, Central African Republic, Chad, Chile, Colombia, Costa Rica, Dominican Republic, Ecuador, Egypt, El Salvador, Ethiopia, Ghana, Greece, Guatemala, Guinea, Haiti, Honduras, India, Indonesia, Ivory Coast, Jamaica,

FIGURE 6.1
Translating Theory into Variables

From York W. Bradshaw, "Urbanization and Underdevelopment: A Global Study of Modernization, Urban Bias, and Economic Dependency," *American Sociological Review* 52 (April 1987): 228. Used with permission.

URBANIZATION AND UNDERDEVELOPMENT 231

Table 2. Relative Urbanization Residuals, Urban Population, and Level of Economic Development

	1960			1980		
Country	RES	URB	GNPC	RES	URB	GNPC
Large positive residuals						
Chile	28.64	67.8	670.9	26.50	80.6	2290.0
Peru	11.12	46.3	520.9	22.59	65.2	1080.0
Colombia	19.26	48.2	350.7	18.73	63.7	1260.0
Zaire	5.84	15.7	104.5	17.18	34.0	200.0
Nicaragua	10.38	41.4	400.1	16.07	53.3	760.0
Mexico	10.69	50.8	712.6	14.82	66.7	1980.0
Brazil	11.39	46.1	505.8	14.39	67.6	2160.0
Bolivia	4.12	24.0	197.3	11.57	44.4	570.0
Egypt	16.18	37.9	221.7	11.51	43.8	550.0
Chad	−5.59	7.0	124.2	10.12	17.8	110.0
Large negative residuals						
Portugal	−20.75	22.5	869.5	−23.57	30.6	2300.0
Thailand	−5.10	12.5	170.7	−20.91	14.4	670.0
P. New Guinea	−22.18	2.7	271.0	−19.73	17.9	780.0
Malaysia	−4.99	25.2	379.5	−19.03	29.4	1580.0
Nigeria	−4.07	13.1	166.1	−18.90	20.4	870.0
Zimbabwe	−18.84	12.6	411.0	−14.63	23.0	780.0
Rwanda	−6.58	2.4	98.8	−13.98	4.3	220.0
Yugoslavia	−11.01	27.9	660.2	−13.39	42.3	2540.0
Kenya	−7.28	7.4	141.9	−12.83	14.2	390.0
Niger	−9.18	5.8	144.6	−10.52	12.5	300.0

Note: RES = relative urbanization residual; URB = percentage of the total population living in urban areas; GNPC = gross national product per capita in U.S. dollars.

worker in nonagriculture (i.e., the percentage contribution of nonagriculture to total gross domestic product (GDP) divided by the percentage of the labor force in nonagriculture) to output per worker in agriculture (i.e., the percentage contribution of agriculture to total GDP divided by the percentage of the labor force in agriculture):

$$\left(\frac{\text{Percent Nonag. GDP}}{\text{Percent Nonag. Labor}}\right) \Big/ \left(\frac{\text{Percent Ag. GDP}}{\text{Percent Ag. Labor}}\right).$$

This indicator is calculated circa 1960 (DISP60). A nation has a high score on this variable when its nonagricultural sector is more productive than its agricultural sector.[11]

Structure of investment dependence. This analysis uses three different measures of external investment: (1) the *total* stock of direct, private foreign investment in 1967 divided by total GNP in the same year (INV67); (2) the stock of direct, private foreign investment in *agriculture* in 1967 divided by total GNP (AINV67); and (3) the stock of direct, private foreign investment in *manufacturing* in 1967 divided by total GNP (MINV67).[12] The investment data are standardized on GNP so that foreign investment is measured relative to the size of a nation's economy (see Delacroix and Ragin 1981; Bradshaw 1985a). Ideally, the investment variables would be constructed closer to 1960. Unfortunately, however, data on the stock of foreign investment are not available for a broad sample prior to 1967.

Concentration of agricultural exports. The concentration of agricultural exports is measured as the percentage contribution of the top three agricultural exports to total agricultural exports circa 1962 (AGCN62). Agricultural exports include all export commodities in Standard International Trade Classification (SITC) sections 0, 1, 2, and 4. These categories include all primary commodities comprised of food and beverages and of nonfood agricultural products. Typically, Third

[11] If the ratio is greater than one, then it indicates that domestic capital is allocated in a way that will maintain greater output per worker in nonagriculture than in agriculture. Unfortunately, this hypothesis cannot be tested because data on the sectoral distribution of domestic investment are not available for a broad sample of nations in 1960. Nonetheless, the ratio of output per worker in nonagriculture to output per worker in agriculture provides an indication of the disparity between these two sectors. In this sample the mean production disparity is 6.50, indicating that nonagricultural workers are six and a half times more productive than agricultural laborers.

[12] While a very small amount of foreign investment in manufacturing may go to rural areas, it is safe to assume that the overwhelming portion of such investment is allocated to urban regions. It should also be noted that other studies of development have used a measure of foreign investment in manufacturing, but these studies have not focussed on urbanization (see Bornschier 1981; Bornschier and Chase-Dunn 1985).

FIGURE 6.2

The Unequal Relationship Between Prose and Figure

Source: See figure 6.1. 231. Used with Permission.

URBANIZATION AND UNDERDEVELOPMENT 237

force also substantiate dependency arguments, as change in level of service employment has a negative impact on rate of economic growth. This finding suggests that Third World service and informal labor may subsidize formal-sector activity throughout the international capitalist system, impeding peripheral economic expansion. The final noteworthy result supports dependency scholars. Specifically, the concentration of agricultural exports has a negative impact on economic growth, suggesting that underdeveloped countries should diversify their agricultural export structure. A lower concentration of such exports would make nations—and their farmers—less vulnerable to low international prices and to other conditions which adversely affect agricultural products that are exported.

What do these findings tell us about the study of urbanization and underdevelopment in the world system? More than anything else, they underscore the importance of testing various theories of urbanization and underdevelopment. This paper showed that two theories help explain the causes of relative urbanization, and three theories help explain the relationship between relative urbanization and economic growth. Failure to test all three views simultaneously would have resulted in incomplete and even spurious findings. For example, some previous research on Third World urbanization has found that modernization theory provides a valid explanation of city expansion. Unfortunately, however, these studies do not test arguments relevant to urban bias and economic dependency. If future studies of urbanization and underdevelopment are going to be useful, then they must transcend current theoretical and ideological particularism.

APPENDIX

Variables	Sources
GNP per capita, 1960	Ballmer-Cao and Scheidegger (1979); World Bank (1976a)
GNP per capita, 1980	World Bank (1983)
Relative urbanization, 1960	Level of urbanization (World Bank 1984); GNP per capita (Ballmer-Cao and Scheidegger 1979; World Bank 1976a)
Relative urbanization, 1980	Level of urbanization (World Bank 1984); GNP per capita (World Bank 1983)
Relative service employment, 1960	Level of service employment (World Bank 1984); GNP per capita (Ballmer-Cao and Scheidegger 1979; World Bank 1976a)
Relative service employment, 1980	Level of service employment (World Bank 1984); GNP per capita (World Bank 1983)
Industrial employment, 1960	World Bank (1984)
Rural-urban disparity, 1960	GDP by industrial origin (World Bank 1976b, 1980, 1984); Composition of the labor force (World Bank 1984)
All investment variables, 1967	Investment capital (Ballmer-Cao and Scheidegger 1979); Total GNP (World Bank 1969)
Concentration of agricultural exports, 1962	United Nations (1963–68); Steinberg (1963) (the latter was used only for data on Haiti)
Gross domestic investment, 1960	World Bank (1984)
Total population, 1960, 1980	World Bank (1984)

REFERENCES

Arrighi, Giovanni. 1973. "International Corporations, Labor Aristocracies, and Economic Development in Tropical Africa." Pp. 105–51 in *Essays on the Political Economy of Africa*, edited by Giovanni Arrighi and John Saul. New York: Monthly Review Press.

Ballmer-Cao, Thanh-Huyen, and Jurg Scheidegger. 1979. *A Compendium of Data for World-System Analyses*. Volker Bornschier and Peter Heintz (eds.). Zurich: Sociological Institute of the University of Zurich.

Bates, Robert. 1981. *Markets and States in Tropical Africa: The Political Basis of Agricultural Policies*. Berkeley: University of California Press.

Berk, Richard. 1983. "An Introduction to Sample Selection Bias in Sociological Data." *American Sociological Review* 48:386–98.

Berliner, Joseph. 1977. "Internal Migration: A Comparative Disciplinary View." Pp. 443–61 in *Internal Migration: A Comparative Perspective*, edited by Alan Brown and Egon Neuberger. New York: Academic Press.

Bollen, Kenneth and Sally Ward. 1979. "Ratio Variables in Aggregate Data Analysis: Their Uses, Problems and Alternatives." *Sociological Methods and Research* 7:431–50.

Bornschier, Volker. 1981. "Dependent Industrialization in the World Economy: Some Comments and Results Concerning a Recent Debate." *Journal of Conflict Resolution* 25:371–400.

Bornschier, Volker and Christopher Chase-Dunn. 1985. *Transnational Corporations and Underdevelopment*. New York: Praeger.

Bornschier, Volker, Christopher Chase-Dunn, and Richard Rubinson. 1978. "Cross-national Evidence of the Effects of Foreign Investment and Aid on Economic Growth and Inequality: A Survey of Findings and a Reanalysis." *American Journal of Sociology* 84:651–83.

Bradshaw, York. 1985a. "Dependent Development in Black Africa: A Cross-national Study." *American Sociological Review* 50:195–207.

______. 1985b. "Overurbanization and Underdevelop-

FIGURE 6.3

Prose Eclipsed by the Texts of Appendix and References

Source: See figure 6.1. 237. Used with Permission.

variables treated in the analysis. This appendix appears to be a table of sorts. And beneath the tabular appendix, the "references" begin, only to end fully three pages later. The references are not prose but figure; they do not talk but only support the writing in serial strings of alphabetized authority. While it makes common literary sense to supply the reader with the names, titles, and dates of material referred to in the body of the text, these are only passing references undeveloped in dialogical argument. The references cited in the text, especially in the literature review section, give journal science its first purchase on disciplinary legitimacy and help ground the particular analysis in a body of common scholarship. But references also serve gesturally throughout the text and especially at the end in their serial array to replace prose, ever the purpose of figural expression in journal science.

Like tables, equations, graphs, and number, the references occupy space, thus diminishing real intellectual labor;[27] in addition, they indirectly serve the purpose of every instance of textual figure in constituting a text beyond itself through which pure representation is achieved. This is not to say that science believes that the world is shaped like a reference page but that figure itself becomes the prose of the world—epistemological constitution proceeding directly through figural gesture and not mediately through prose. The preponderant reference section gives further legitimacy to the other instances of figural expression throughout the text, especially tables, number, and equations. All these figures overwhelm mediate prose, making way for a text seemingly disclosing the world directly, without recourse to the labored intelligence of thought.

All the instances of figure I have cited here both diminish thought and heighten the text's science aura. They become text to prose's subtext; people are led to read all the figured marginalia of journal science before they delve into the prose. Prose becomes the margin, figure the center, thus losing its constitutive salience for the argument. Of course, science does not argue but only represents, or so it would lead us to believe. I am saying that these marginal gestures are really central literary constructions in their own right; far from being ancillary to the argument, they are text, whereas text is increasingly whittled down, itself reduced to thick technical jargon scarcely differentiable from what surrounds it in the way of figure.

Science, through the extravagant use of figure, causes the rate of prose to fall further. Again, the solution is not more prose and less figure; a ban on graven images solves nothing, especially when prose itself can be technified, mechanized, moronized. Humanists would thus solve world crises by adding poetry. But instead we must read what would pass for mere technique, gesture, figure as poetic—indeed, political—texts in their own right. They embody passion and perspective, and they recommend certain states of affairs. Figure is an argument for a world like itself—patterned, mathematized, monetarized, deauthorized. Figure freezes a world from the inside already nearly frozen from the outside. It adds momentum to what is already happening in the way of political economy and culture. I do not demonize the literary (and thus political) strategies of positivism; journal writing only intensifies the falling rate of intelligence, thus checking the falling rate of profit.

Journal science is not a *deus ex machina*. In fact, nothing is; everything relates, adding value to the world's irrational totality. The latter-day embodiment of positivism in discipline, journal science is only a recent excrescence of the will to power. This sounds portentous both to everyday science authors and literary critics. Surely things are not as dramatic as all that. But my deconstructive work here presupposes and is in turn presupposed by a theory of totality within which small literary gestures refract the dehumanization of the world in larger terms. Indeed, the more things fit together in a concatenated totality of discipline, the more decentered they appear. Differentiation seems to be the paramount principle of modernity, governing both material and literary practices.[28]

Many would argue that one must restrict critiques such as this to local discourses. Yet people like Adorno,[29] Habermas[30] and Foucault[31] open plural discourses to a common reading of their disciplining tendencies. This book contributes in its small way to that effort; indeed, I read in science journals in order to reconstruct a literary logic common to all discourses in which authorial subjectivity is buried underneath the requirements of administration. Journal science is no less a part of contemporary culture (and thus reproductive of it) than Hollywood, Madison Avenue, or music. Although science would claim an exemption from deconstructive readings of this sort, we must refuse to grant it that exemption. Instead, its discursive strategies must be interrogated politically and not simply dismissed as literary folkways.

Nothing is unstudied in journal science. Yet this does not imply elaborate design or even cupidity on the part of authors and editors as they subtly subordinate text to subtext. Writing is contingent; like any material practice, it is done through the filter of history. As such, it only partly understands itself. Some gestures are purposeful, whereas others simply carry over from the past. The more we understand the strategic nature of our expressive work, the more control we have over it—the better it suits our larger rhetorical and thus political purposes. Positivism is alone among literatures in that it seems oblivious to the way it reposes on the page and thus inserts itself between past and future.[32] Fiction knows its artifice well; it celebrates authorial imagination as an epistemological resource.[33] That is why we say that the best fiction is the most truthful; circularly, the most truthful is the best. Positivism suppresses its narrativeness precisely to strategize covertly. In the process, in appearing oblivious to its rhetorical nature, it seduces readers and other writers into its political metaphysic.

I want to understand the figural work of science this way. Science writers surround prose with figure and interpose figure between fragments of prose, and even figure prose itself, for political reasons. They add momentum to a certain history in which people read readings of the world uncritically, failing to recognize and thus engage with their constitutionality. Figure not only diminishes prose (and thus the possibility of mediation) but also suggests a figured world in which things fit together neatly—caused, patterned, counted, unidimensional. Representation constitutes. Where figure takes over the constitutional role of representation from prose (and prose becomes a kind of figuring itself), it suggests a world to which it bears the secret affiliation of metaphysical support. Even description recommends—the more so, the more it conceals its rhetorical, partisan quality. So-called value freedom enslaves one to value.

It is immaterial whether scientists know this or not about their literary work. Inasmuch as oppression reduces self-consciousness, it is unlikely that people who perpetuate it really understand what is going on. At the same time, we cannot assume absolute unconsciousness; that is structuralism. Structures exist, enforcing unconsciousness. Yet unconsciousness can undo itself. That is the riddle of history. Opportunities are never closed; that is why people write, struggle, hope. In writing, they also ironically close those opportunities when they portray an inert order of things, as positivism does. The key is not whether writers, in perpetuating a figured world by figuring it, have a coherent strategy; their politics is covert, perhaps unknown even to them. Factuality wins the day when its representation is seen to be sheer reflection, not also constitution. Therefore, the people who control journal science and thus reproduce it in its image may not understand their own contribution to the totality; for them, the figuring of science is its essence (in the strongest sense) or merely a binding norm (in the weaker sense). They do not necessarily understand their own writing deconstructively as the political work it is—freezing an already chilly social universe, figuring a figural world. A few do; most do not. Yet that is beside the point. Science can be read, hence rewritten.

Chapter 7 Science Aura III: Findings Figured

Table

Figure is science's gestural work, carried out in the margins of the page as well as in its midst. Even prose can virtually become figural when enough technical jargon, acronym, and number are packed into it. Number is a crucial kind of figural work, enhancing the science aura directly (a subject for the following section of this chapter). Here I want to consider the various literary ways in which science writes without words, or at least without prose having any sort of presence in the argument. I have already given examples of equations and tables. Let me expand this discussion. Most methodologists are interested in the logics of inference, sampling, and causal modeling in what they take to be purely technical terms. But I insist that no "purely technical terms" govern the resolution of epistemological debates; there is only rhetoric. One hopes that the most persuasive argument is the best one; that is the risk run in a democracy—Habermas's ideal speech situation. Methodologists do not want to run that risk when they hold out what they contend are purely technical criteria with which to solve science's problems. This is why so much of the space in journal science is taken up with methodological quibbles, elaborations, tangents.

Instead, in looking at methodology as a literary way of making arguments, I reject the notion of an inherently technical reason.[1] Methodology is art work, rhetorically styled to form consensus. As such, then, it is political work. Methodology wants to persuade others of its view of the world. Positivist method does so by portraying itself *as* a view of the world, if not a worldview. It supposes that representation is not constitution. It also suggests itself directly as a depiction of the order of things—a figured, tabled, numbered world. Once we accept figure as a window on things, we will more readily believe that those things could not be different. Figural work in journal science freezes history in the frame of a photographic realism. In this and the next four sections of the chapter, I discuss the modalities of figural work in aesthetic and thus political terms. I consider in turn the use of tables, equations, graphs, causal models, and picture.

One might object that no one believes that the figures in journals science look exactly like the world. No. But the figured, tabled, numbered world is thought to disclose itself presuppositionlessly to the lens of high science. This assumes that photography itself is merely a technical vehicle of communication; many, such as Susan Sontag,[2] argue convincingly against that view. Figural readers quickly figure out that the figured, tabled, and numbered world in its presumed lawfulness is better disclosed gesturally than through slow prose. Figure moves more quickly from page to mind through retina. Prose must be thought, reconstituted, evaluated. Instantaneity rules in fast capitalism, disqualifying mediation as a prefigural artifact. Journal science in its elaborate figural display hopes to convince readers that whereas the tabled, numbered world is not exactly the same as the tables and numbers themselves, at least they are better, clearer approximations than old- age prose. The rate of prose falls in direct relation to the rate of overall intelligence.[3]

Perhaps I seem to suggest a golden age when prose ruled and figure knew its place. I do not. In fact, prose can be misleading, ontologizing, deadening. And prose can be figured—jargonized, numbered, disciplined—so that it scarcely differs from the figural work surrounding it. Although there was no golden age of wordsmiths, the present age is losing value quickly when communication is increasingly reduced to technique. In itself, figure can be highly evocative and imaginative. Yet science intends figure as mere representation; while tables are less definitively descriptive than photographs themselves, they are intended to be more descriptive than floral prose. If prose never knew a golden age, then figure is not utopia either. The crucial issue is that literary work, whether word or drawing, is disguised as a presuppositionless representation of a fixed world.

In beginning my discussion of the use of figure, I have shown some tables that occupied nearly half the page of journal science. Figure 7.1 is a case where they occupy much more. I want to use this as an example to suggest how the literary analysis of figure in journal science can proceed. I then elaborate my analysis in discussing the variety of figurings covering the pages of journal science. In the analysis, one must attend to a variety of texts in the literary use of figure.[4] Keep in mind that I reject the dichotomy of text and margin when the text is increasingly dominated by figure and figured prose, and the margin becomes prose itself. One of the most important political insights of Derridean deconstruction is that in an ideologized world we can frequently learn the most by examining not what one might take to be the center but by concentrating on marginalia as the main text in their own right.[5] Derrida unfortunately makes this a methodological principle, dooming deconstruction to political irrelevance in the fetishes of literary criticism.[6] But he is right to suggest that we might examine distinctions presenting themselves to us as simple differentiation; the hierarchies they conceal as well as overturn are presented to us as the opposite of what they really suggest about the priority of their terms. Thus, when science would lead us to think that word and figure are fundamentally discontinuous discourses

IMPACT OF GENDER SEGREGATION 581

Table 1. Descriptive Statistics and Analysis of Variance for Variables by Gender Mix of Men's Work Setting

		Gender Mix[a]					
Variables	Total Sample ($N=822$)	Male ($N=416$)	Mixed-but-segregated ($N=131$)	Predominantly Male ($N=196$)	Mixed ($N=59$)	Female ($N=20$)	η^2
Job satisfaction	37.67	38.05	38.45	38.88	35.80	38.00	.57
	(10.06)	(9.96)	(9.95)	(10.12)	(10.71)	(9.88)	
Job depression	19.62	19.36	19.50	20.11	20.57	18.17	.61
	(5.88)	(5.78)	(5.56)	(5.84)	(8.75)	(5.88)	
Job-related self-esteem	59.62	60.15	60.22	58.62	57.96	59.05	.67
	(9.75)	(9.45)	(9.59)	(10.23)	(9.02)	(13.57)	
Race	.91	.92	.90	.90	.88	.90	.31
Age	37.70	37.75	35.92	38.29	39.32	37.50	.47
	(12.87)	(12.72)	(13.01)	(12.57)	(13.95)	(14.58)	
Squared term for age[b]	165.46	161.57	170.57	157.76	194.60	202.05	.36
	(183.05)	(176.82)	(177.08)	(181.80)	(188.89)	(305.50)	
High school graduate	.36	.41	.29	.33	.27	.40	2.19**
Some college	.20	.16	.21	.26	.25	.20	1.04*
College graduate	.18	.12	.34	.17	.29	.25	4.81*
Unmarried	.17	.14	.27	.11	.29	.35	3.23***
Married to housewife	.57	.63	.38	.60	.54	.40	3.46***
Wife works	.26	.23	.35	.29	.17	.25	1.35**
Lives in south	.32	.36	.29	.28	.24	.40	1.00*
Training time (SVP)	5.84	6.01	5.89	5.50	5.98	4.63	1.90***
	(2.05)	(1.97)	(2.03)	(2.20)	(1.82)	(2.14)	
Danger	.51	.63	.40	.42	.20	.40	7.13***
Physical demands	2.57	2.72	2.37	2.45	2.31	2.55	2.41***
	(1.05)	(1.02)	(1.12)	(1.05)	(1.00)	(1.19)	
Craft union	.13	.24	.02	.03	.02	.00	10.00***
Other union	.26	.17	.35	.30	.48	.35	4.79***
Job tenure	4.61	4.67	4.51	4.53	4.67	4.65	.17
	(1.82)	(1.85)	(1.74)	(1.78)	(1.90)	(1.76)	
Employer tenure	5.14	5.15	4.98	5.21	5.27	5.00	.20
	(1.88)	(1.92)	(1.83)	(1.85)	(1.92)	(1.89)	
Steady	.92	.87	.95	.97	.98	.90	3.45***
Hours worked per week	44.75	45.07	43.72	45.26	44.01	42.10	.54
	(9.56)	(9.33)	(9.46)	(10.43)	(8.33)	(9.19)	
Organizational scale	3.45	3.21	3.88	3.56	3.61	3.89	1.76***
	(1.95)	(1.92)	(2.00)	(2.00)	(1.87)	(2.16)	
Autonomy	3.22	3.18	3.35	3.20	3.24	3.20	.39
	(.94)	(.94)	(.86)	(.96)	(.98)	(.95)	
Authority	.48	.46	.50	.49	.56	.45	.30
Log earnings	9.16	9.19	9.10	9.15	9.24	8.95	1.08*
	(.49)	(.47)	(.51)	(.50)	(.53)	(.60)	
Log spouse earnings[c]	2.43	1.98	3.66	2.61	2.08	2.99	2.48***
	(3.80)	(3.55)	(4.30)	(3.94)	(3.76)	(4.37)	
Occupational status	41.28	36.42	48.35	44.33	50.38	39.47	5.79***
	(23.32)	(21.32)	(25.16)	(23.90)	(22.35)	(20.59)	

Note: Standard deviations shown in parentheses, except for dichotomous variables. η^2 is the percentage of the total variation that exists across the five gender mix categories. Asterisks indicate that the mean differences among the five groups are statistically significant at the * $p \leq .05$, ** $p < .05$, or *** $p \leq .01$ level (two-tailed test).

[a] See text for definition of categories.

[b] Squared deviation from average age in sample.

[c] Equals zero for men who do not have a working wife.

related depression and job-related self-esteem, with men in male-dominated and female-dominated settings expressing the greatest psychological well-being.

These results are intriguing. Men in predominantly female work settings regard their situation almost as favorably as men in all-male ones, despite significantly fewer extrinsic rewards. Men in female-dominated settings exhibit some characteristics often regarded as typical of women's orientation to work; strong job satisfaction notwithstanding low pay and prestige. This may reflect a rationalization process whereby workers elevate perceived satisfaction, self-esteem, and the like to compensate for the low pay of their jobs, consistent with cognitive theories of "insufficient justification" and its effects on attitudes (Aronson 1972, chap.

FIGURE 7.1
The Table Dominates the Page

From Amy S. Wharton and James N. Baron, "So Happy Together? The Impact of Gender Segregation on Men at Work," *American Sociological Review* 52 (October 1987): 581. Used with permission.

we might read word as figure and figure as word. And when science suggests to us a priority of its proseful text over the embellishing figural subtext, we might interrogate that priority for what it in fact conceals about the real state of affairs. The assumption behind this sort of deconstructive work is that domination dominates and thus transforms the very distinctions and hierarchies by which we order our ordinary sense-making practices.

Thus, methodologically, we attend not to the busy calculations represented in the table reprinted in figure 7.1 or to the pattern of data reported. The table wants us to read it that way. Instead, we look at the semiotic nature of the table (e.g., its ratio of number to prose), as well as at the relationship between the table and the prose text beneath it. Already we know something interesting: The prose reposes under a vast table literally squashing it. The table is more important than the prose barely visible beneath it. Also, the table is primarily numbers—mean scores along certain variables, standard deviations from the mean, and percentages of total variation across the five categories of "gender mix." The variables are described in words at the far left-hand side of the table.

Figure adds up in journal science. It is important to understand both its relation to other semiotic media on a single page and its frequency. Figure increasingly dominates journal science. And the internal relations between prose and figure change even within single articles. Everything matters—the use of figure and number, the dominance of figure on the page, its frequency, its many natures. All are literary gestures; together they constitute a text whose fabric as well as letter can be read. Altogether too much time is spent reading the wrong text in journal science; while the words and data matter, they are less and less the whole story. They are not even the real story. Science, in gesturing a world, politically freezes the present world into the tables, patterns, and numbers with which it addresses its subject semiotically. The medium is the message where nondiscursive discourses are concerned. They disclose a world—and a conception of good and bad—in the very terms with which they attempt to reflect the real. Representation necessarily represents itself—a world to which it bears a willingingly subordinate relationship, ever the posture of ideology.

Let me return again to tables, one of the five generic forms of figural work in journal science.[7] A question is how much space they claim on the page; another question is their frequency; further questions are their reliance on number over prose and then the different sorts of number and figure they utilize. Examples can help me elaborate my discussion. Without suggesting a strict logic of hierarchy, let me move from the least figural to the most figural tables.

The table in figure 7.2 is all words. As such, it wants only to suggest some "hypothesized effects" further pursued in the article. The figure helps organize and clarify the hypothetical reasoning in the text. It spells out what

100 AMERICAN SOCIOLOGICAL REVIEW

Table 1. Hypothesized Effects of Beliefs about Devaluation and Discrimination

	First-Treatment Contact Patients	Repeat-Treatment Contact Patients	Formerly Treated Community Residents	Community Untreated Cases	Community Respondents Without Pathology
Distinguishing Features					
Currently in treatment?	yes	yes	no	no	no
Previous treatment?	no	yes	yes	no	no
Level of psychopathology?	high	high	mixed	high	low
Prediction concerning the effect of beliefs about devaluation-discrimination rejection on demoralization within the group.	Significant effect expected	Significant effect expected	Weak effect expected—residue of past labeling	No effect expected	No effect expected
Prediction concerning the effect of beliefs about devaluation-discrimination on employment and earned income.	No effect expected—no past history of a label	Significant effect expected	Significant effect expected but a smaller one than among repeat contact patients currently in treatment	No effect expected	No effect expected

psychiatric patients (N = 164) in the Washington Heights section of New York City between 1980 and 1983 (Dohrenwend, Shrout, Link, Martin, and Skodol 1986). The community respondents were obtained by following up a sample collected approximately six months earlier. In the original sample, households were enumerated through telephone directories and contacted to see if an eligible respondent between 19 and 59 years of age lived there. In 93 percent of the households, screening information was provided; 68 percent of these contained one or more eligible respondents. The goals of the original study were methodological, involving random assignment of respondents to two versions (one-month versus one-year recall of symptoms) of the Psychiatric Epidemiology Research Interview (PERI). Given this focus, the major effort was placed on testing the different forms of the PERI in contrasting ethnic and socioeconomic groups. Intensive efforts to interview hard-to-schedule respondents and to convert refusals were not undertaken initially. Later, when the substantive study was implemented toward the end of the methodological study, provision was made for checking on a subsample of "hard-to-schedule" subjects (N = 48).[1] In the end, 57 percent of the 943 eligible respondents in the original sample were interviewed successfully.

With respect to the representativeness of the sample, I reasoned that the hard-to-schedule respondents would be more like the larger pool of nonrespondents than the more easily obtained individuals. This reasoning seemed appropriate: without extensive effort these individuals would have been nonrespondents. As a result, they provide the best check available on the sample's representativeness. Hard-to-schedule respondents were compared to other community respondents in their opinions about the attitudes of most people toward mental patients. The means for the devaluation-discrimination scale were quite similar for hard-to-schedule and all other community respondents and not significantly different.

In addition to generating a sample of hard-to-schedule respondents, extensive efforts were initiated to ensure an adequate reinterview rate in the second wave of data collection. As a result, 79 percent of the initial sample (429) was located and reinterviewed.

The patient sample was selected from outpatient clinics and inpatient facilities in the same general area of New York City. The goal was to select patients in two diagnostic categories, major depression and schizophrenia and schizo-

[1] This group included 33 individuals who refused initially but were converted with intensive effort; 12 who failed to keep two appointments (our usual cutoff as a lost case) but were obtained on a third or later appointment; and 3 who could not be reached by telephone after eight attempts at different times of day (our usual cutoff as a lost case) but were reached on a ninth or subsequent attempt.

FIGURE 7.2
Figure Comprised of Words

From Bruce G. Link, "Understanding Labeling Effects in the Area of Mental Disorders: An Assessment of the Effects of Expectations of Rejection," *American Sociological Review* 52 (February 1987): 100. Used with permission.

the authors might expect to find in their research. Interestingly, it also occupies less than half the page. Like most tables that must share space with prose, this one perches at the top of the page. Rarely are tables found at the bottom of the page or even in the middle. They begin the page's text, thus inherently commanding early attention from the reader.

This prose figure clearly reveals its irony; prose is chopped into pieces and then ordered figurally. Syntax dissolves, as does mediating reasoning. In itself this suggests that the main body of the prose text is essentially ancillary to this pithy summary; these are the variables, these are the issues, this is what we expect to find. One can skip the slow prose of the text; this table appears to argue economically, preparing the way for data analysis. The prose figure accelerates the rates of writing and reception. The table as figure lacks number, one of the most frequent forms of figure. It anticipates numbered presentation later on.

Figure 7.3 is more typical. Here the table takes up a whole page, as many tables do. It combines prose and number. The prose is the names of independent and dependent variables; the numbers summarize both money and certain statistical coefficients. Numbers predominate; prose is marginal. As well, there is some empty space in the midst of the table where no number values are entered. Underneath the table we are told the data source and offered other clarifications about the statistics in question. This is the sort of table every journal reader would master readily; it is the standard canvas of science. The table is printed in the midst of the findings section, adding crucial detail to the prose discussion.

The table in figure 7.4 is similar. This table, too, occupies the whole page. It reports statistical measures of the relationships between dependent variable and independent variables. Again, it is supported with some statistical clarifications underneath it. The variables are named in prose. While basically similar, the table is busier than the one just above it. There is less unoccupied space. As well, because the table would not fit lengthwise on one page it is split into two with one half resting below the other. Again, interpretation is unproblematic for aficionados. For outsiders or recent initiates, working through the dense layers of figure is arduous and time-consuming.

One wonders, thus, if the economic rationale for figural expression is convincing. Numbers are as semiotic as words; they construct the world and they must be constructed in turn. But science believes that these tables afford direct access to the world's mysteries; as numbered and patterned intelligence, they disclose much more than mere words. Tabular presentation of numbered data is a prerequisite of all science, even if the analysis does not need it. Articles using speculative approaches are numbered and tabled in order to fulfill a binding norm of journal science—no matter that the data are as epigrammatic as words, frequently as opaque. Tabled and

Table 3. Effects of Human Capital, Race/Ethnicity, Industry Shift, Occupational Mix, and Gender Composition on 1979 Dollar Earnings by Sex (Beta Coefficients in Parentheses)

	Women's Earnings				Men's Earnings			
Independent Variables	(1)	(2)	(3)	(4)	(1)	(2)	(3)	(4)
Age	381** (0.809)	385** (0.818)	373** (0.792)	378** (0.804)	732** (0.968)	732** (0.968)	696** (0.920)	719** (0.951)
Age^2	−4** (−0.670)	−4** (−0.672)	−4** (−0.648)	−4** (−0.658)	−7** (−0.705)	−7** (−0.705)	−6** (−0.658)	−7** (−0.691)
Education	240** (0.143)	224** (0.133)	226** (0.134)	222** (0.132)	409** (0.169)	406** (0.168)	434** (0.180)	417** (0.172)
High school graduate	614** (0.050)	518** (0.042)	634** (0.052)	517** (0.042)	672* (0.035)	657* (0.035)	643* (0.034)	499* (0.026)
College graduate	2,452** (0.134)	2,556** (0.139)	2,456** (0.134)	2,459** (0.134)	3,537** (0.122)	3,517** (0.121)	3,796** (0.131)	3,184** (0.110)
English proficiency	908** (0.053)	718** (0.042)	821** (0.048)	717** (0.042)	1,848** (0.070)	1,848** (0.070)	1,795** (0.068)	1,783** (0.068)
Hispanic	107 (0.009)	130 (0.010)	145 (0.012)	147 (0.012)	−2,100** (−0.100)	−2,109** (−0.100)	−1,908** (−0.091)	−1,938** (−0.092)
Black	233† (0.014)	298* (0.018)	351* (0.022)	371* (0.023)	−3,116** (−0.104)	−3,112** (−0.104)	−2,852** (−0.095)	−2,835** (−0.095)
Occupational mix[a]		713** (0.051)	633** (0.045)	417** (0.030)		1,681** (0.031)	2,291** (0.042)	384 (0.007)
Industrial shift[a]		738** (0.070)	655** (0.062)	675** (0.064)		−297 (−0.008)	24 (0.001)	−118 (−0.003)
Proportion female, 1980			−1,829** (−0.077)	—[b]			−4,562** (−0.122)	—[b]
Feminization rate			—[b]	1,408** (0.075)			—[b]	2,728** (0.125)
$\hat{\lambda}$	1,208** (0.048)	1,136** (0.045)	1,067** (0.043)	1,076** (0.043)	−38** (−0.055)	−38** (−0.054)	−38** (−0.055)	−37** (−0.053)
Constant	−7,510	−7,034	−5,819	−7,094	−12,867	−12,731	−10,948	−13,486
R^2	.279	.285	.291	.290	.344	.345	.359	.360
F	514.4**	459.7**	443.0**	441.6**	450.9**	392.5**	391.0**	392.0**

Source: U.S. Bureau of the Census, 1970 and 1980 Public Use Microdata Samples.
Note: Net of controls for nativity and marital headship and full-time employment status.
[a] Defined in Appendix B. Note that the structural change terms are based on gender-specific analyses for like sex.
[b] Not included in equation.
* $p<.05$.
** $p<.001$.
† $p<.10$.

FIGURE 7.3

Number Squeezes Out Prose in Tables

From Marta Tienda, Shelley A. Smith, and Vilma Ortiz, "Industrial Restructuring, Gender Segregation, and Sex Differences in Earnings," *American Sociological Review* 52 (April 1987): 201. Used with permission.

Table 3. Regression of Minutes Together in Various Activities on Family Workday, Life Cycle, and Sociocultural **Measures** (*b*/standardized *b* in parentheses)

	Dependent Variable: Time Together							
	Any Activity		Child Care[a]		Fun		Homemaking	
Independent Variables	Husbands' Account	Wives' Account	Husbands' Account	Wives' Account	Husbands' Account	Wives' Account	Husbands' Account	Wives' Account
Total minutes of work (husband + wife)	−0.151*	−0.115*	−0.002	−0.011**	−0.037*	−0.022	−0.018**	−0.060*
	(−0.239)	(−0.187)	(−0.032)	(−0.140)	(−0.158)	(−0.118)	(−0.151)	(−0.266)
Off-scheduling (minutes)	−0.190*	−0.225*	−0.004	−0.010**	−0.030**	−0.022	−0.020*	−0.048*
	(−0.309)	(−0.374)	(−0.089)	(−0.148)	(−0.129)	(−0.123)	(−0.171)	(−0.219)
Family income (1000s)	0.462	0.317	−0.186**	−0.053	0.003	0.084	0.080	−0.098
	(0.043)	(0.030)	(−0.173)	(0.035)	(0.001)	(0.026)	(0.040)	(−0.026)
Family prestige	−1.447	−2.148*	0.208**	−0.142	0.968*	0.057	−0.134	−0.414
	(−0.102)	(−0.156)	(0.162)	(−0.080)	(0.183)	(0.014)	(−0.051)	(−0.083)
Respondent race (0 = nonwhite, 1 = white)	68.639	52.739	3.061	−9.27	13.287	10.787	7.502	1.365
	(0.072)	(0.057)	(0.041)	(−0.090)	(0.037)	(0.038)	(0.042)	(0.004)
Number children in household	−3.891	−5.439	−1.790	1.723	0.145	−1.40	−1.736	−4.346
	(−0.032)	(−0.046)	(−0.143)	(0.099)	(0.003)	(−0.039)	(−0.077)	(−0.101)
Preschoolers? (0 = no, 1 = yes)	−7.959	8.100	9.794*	9.870*	−10.092	−0.503	−0.707	1.190
	(−0.022)	(0.023)	(0.337)	(0.244)	(−0.075)	(−0.005)	(−0.011)	(0.009)
Years in present marriage	2.710**	2.530**	−0.049	−0.062	0.194	0.617	−0.044	0.812
	(0.159)	(0.152)	(−0.029)	(−0.027)	(0.003)	(0.121)	(−0.014)	(0.134)
Constant	374.958	413.016	1.762	33.00	11.087	33.392	43.106	132.911
N	177	177	141	141	177	177	177	177
R^2	0.213	0.224	0.184	0.126	0.100	0.058	0.061	0.153
Significance (*p* value)	0.000*	0.000*	0.001*	0.019*	0.025*	0.246	0.217	0.000*

	Dependent Variable: Time Together							
	Service-Help		Talking[a]		Watching TV		Meals	
Independent Variables	Husbands' Account	Wives' Account	Husbands' Account	Wives' Account	Husbands' Account	Wives' Account	Husbands' Account	Wives' Account
Total minutes of work (husband + wife)	−0.010	−0.016*	−0.013**	−0.029*	−0.050*	−0.023	0.039*	0.039*
	(−0.117)	(−0.169)	(−0.132)	(−0.297)	(−0.166)	(−0.102)	(−0.245)	(−0.272)
Off-scheduling (minutes	−0.015*	−0.014*	−0.012**	−0.012**	−0.053*	−0.039*	−0.028*	−0.029*
	(−0.182)	(−0.154)	(−0.132)	(−0.128)	(0.180)	(−0.178)	(−0.179)	(−0.207)
Family income (1000s)	0.326*	0.174	−0.068	0.163	0.113	0.481	0.175	0.413*
	(0.224)	(0.108)	(−0.042)	(0.099)	(0.022)	(0.125)	(0.064)	(0.168)
Family prestige	−0.375*	−0.434*	0.224	0.094	−1.520*	−0.281	−0.173	−0.130
	(−0.197)	(−0.206)	(0.105)	(0.043)	(−0.227)	(−0.056)	(−0.049)	(−0.040)
Respondent race (0 = nonwhite, 1 = white)	−26.430*	0.214	9.793	1.234	23.784	7.630	13.135	12.977
	(−0.205)	(0.002)	(0.068)	(0.008)	(0.053)	(0.022)	(0.054)	(0.060)
Number children in household	−0.309	−1.386	1.354	0.078	−1.320	3.477	−1.872	−2.358
	(−0.019)	(−0.077)	(0.074)	(0.004)	(−0.023)	(0.080)	(−0.061)	(−0.085)
Preschoolers? (0 = no, 1 = yes)	−2.44	−1.478	−2.119	−2.055	−4.58	1.459	−10.530	1.000
	(−0.050)	(−0.028)	(−0.039)	(−0.037)	(0.003)	(0.011)	(−0.116)	(0.012)
Years in present marriage	−0.500*	−0.195	−0.201	−0.066	1.261**	0.139	0.602**	0.472
	(−0.216)	(−0.076)	(−0.078)	(−0.025)	(0.155)	(0.023)	(0.139)	(0.121)
Constant	68.274	54.079	9.971	35.999	166.675	65.687	78.582	59.608
N	177	177	177	177	177	177	177	177
R^2	0.130	0.073	.062	.112	.142	.055	.178	.185
Significance (*p* value)	0.003*	0.109	.202	.009*	.001*	.284	.000*	.000*

[a] Respondents with children only.
* $p < 0.05$.
** $p < 0.10$.

FIGURE 7.4

The Domination of the Journal Page by Number and Figure

From Paul William Kingston and Steven L. Nock, "Time Together Among Dual-Earner Couples," *American Sociological Review* 52 (June 1987): 397. Used with permission.

numbered reasoning is thought to epitomize the logic of scientific discovery.

I have an additional concern. In usurping the whole page, indeed whole pages, in journal science, these thick tables seem to have a self-evident privilege. It is assumed that they convey vital information. But even a cursory check of the prose discussion indicates that only a small slice of the information is really being interrogated in the analysis. Of course, one might defend an encyclopedic approach to tabular representation on grounds of procedural openness; readers have a right to the whole cloth of science so that they can check the results in their own work. Or one might defend encyclopedism simply on grounds of possible interest; a plurality of readers suggests a plurality of approaches to the findings in question. But there is no self-evident logic of inclusion, especially when science already uses appendixes as a dumping ground for potentially interesting but nonessential constructions. Tables do not possess inalienable rights. As such, they frequently proliferate simply as a norm of journal science; tabled analyses are circularly the most scientific.

As well, as we see further in chapter 9, a full panoply of data satisfies the requirement of grounded interpretation. One must not draw conclusions unsupported by data; to go "beyond" the data is illegitimate according to science's rules of evidence. The more tabular work, the better. One thus wards off the suspicion that one has doctored the data, fudged the analysis, violated professional norms. Even when the tabled data do not enter the analysis itself, it is thought to be somehow satisfying that data occupy such a large place in the text. A tabled analysis is good writing. And one might add that science aura is enhanced when one figures one's work in plural ways—table, equation, graph, causal model, picture. Typically, tables are preceded and followed by other figural gestures, suggesting a real command of methodology—or, as I prefer, certain narrative techniques. The denser the text with a variety of gestures, the more it will read credibly as sheer representation, capturing the world in its heterogeneous cragginess.

Now let me turn to figural sequence in particular. Two consecutive pages in an article sport two different figures, suggesting a logic of development characteristic of good science. We then skip a page and arrive at yet a higher methodological level. The table in figure 7.5 reports simple percentages of men and women with certain economic and educational characteristics. While the table rests heavy on the prose, it does not take up the entire page. And the table is dominated by the prose names of these economic and educational characteristics. The numbers merely add up to about 100, reflecting simple percentaging at the right edge of the table. This is tabular work at its most basic; the numbers are descriptive in the most ordinary sense.

TABLE 1

PERCENTAGE DISTRIBUTIONS OF FATHER'S AND RESPONDENT'S CLASS AND EDUCATION: EMPLOYED PERSONS OF NONFARM ORIGIN, AGES 25–64

Variable	Men	Women
Father's class (*N* = 13,358 men, 6,455 women):		
Capitalists	10	9
Managers	4	3
Other supervisory	3	4
Petty bourgeoisie	15	15
Nonsupervisory workers	67	68
Total	99	99
Father's education (*N* = 12,940 men, 6,260 women):		
No degree	48	50
Primary school leaving certificate	34	33
Vocational degree	6	5
Higher primary	2	2
Baccalauréat or higher technical	5	5
University degree	6	6
Total	101	101
Respondent's education (*N* = 13,981 men, 6,780 women):		
No degree	23	28
Primary school leaving certificate	28	30
Vocational degree	23	19
Higher primary	7	4
Baccalauréat or higher technical	6	4
University degree	13	15
Total	100	100
Respondent's class (*N* = 14,405 men, 6,450 women):		
Capitalists	8	3
Managers	4	1
Other supervisory	16	8
Petty bourgeoisie	6	5
Nonsupervisory workers	66	83
Total	100	100

many sons as daughters of capitalists inherited their father's class position (27% vs. 9%). Similarly, 13% of the sons of managers and 25% of the sons of other supervisors inherited their fathers' class positions compared with only 2% and 13%, respectively, of daughters. The expected movement of offspring of capitalists and petty bourgeoisie into control positions is also greater for men; 22% of the sons of capitalists versus 12% of the daughters became managers or supervisors, and 18% of the sons of petty bourgeoisie versus 9% of the daughters acquired such positions. Roughly the same percentages of sons and daughters of petty bourgeoisie inherited their fathers' classes (13% and 10%), but about three-quarters of the sons of nonsupervisory workers inherited their fathers' classes compared with

FIGURE 7.5
Table Composed of Simple Percentages

From Robert V. Robinson and Maurice A. Garner, "Class Reproduction Among Men and Women in France: Reproduction Theory on Its Home Ground," *American Journal of Sociology* 91, no. 2 (1985): 262. Used with permission.

TABLE 2

PERCENTAGE DISTRIBUTION OF FATHER'S CLASS BY RESPONDENT'S CLASS: EMPLOYED PERSONS OF NONFARM ORIGIN, AGES 25–64

	SON'S CLASS					
FATHER'S CLASS	Capitalist (N = 1,154)	Manager (N = 537)	Other Supervisory (N = 2,094)	Petty Bourgeois (N = 796)	Worker (N = 8,757)	Total (N = 13,338)
Capitalist (N = 1,345)	27	5	17	7	43	99
Manager (N = 554)	7	13	29	4	48	101
Other supervisory (N = 429)	7	7	25	5	56	100
Petty bourgeois (N = 2,018)	13	4	14	13	56	100
Nonsupervisory worker (N = 8,992)	5	3	15	4	73	100
	DAUGHTER'S CLASS					
	Capitalist (N = 204)	Manager (N = 68)	Other Supervisory (N = 478)	Petty Bourgeois (N = 298)	Worker (N = 4,992)	Total (N = 6,040)
Capitalist (N = 537)	9	1	11	7	72	100
Manager (N = 197)	4	2	15	5	75	101
Other supervisory (N = 223)	2	1	13	3	81	100
Petty bourgeois (N = 906)	6	1	8	10	75	100
Nonsupervisory worker (N = 4,177)	2	1	7	4	86	100

FIGURE 7.6
The Creeping Hegemony of Percentages in Tables
Source: See figure 7.5. 263. Used with Permission.

The table in figure 7.6 takes science to a higher level, even if the progression is not obvious to the quick science reader. As such, the authority of the analysis is enhanced. Here the table takes up the entire page. Prose is kept to a minimum; number predominates even though, interestingly, the slim numbers are spread across the page, leaving ample space unoccupied. Over two consecutive pages, the ratio of prose to figure has been turned upside down. In the table in figure 7.5, words held sway; in the table in figure 7.6 numbers constitute the text. Still, we are dealing with rudimentary percentage distributions, nothing fancy.

We then skip a page. That page represents a crucial transition from crude to sophisticated number work. The table in figure 7.7 reports results of certain statistical manipulations as a way of summarizing causal relations among variables. This is much more advanced than mere percentages in the sense that it suggests causality. Ironically, though, these tabled data gain in supposed representational efficacy what they lose in naive one-to-one correspondence to the world. Whereas percentages summarize simple observable characteristics, the correlations reported in the table in figure 7.7 reconstitute the data through mathematical reasoning; one cannot "see" causality. Statistics represent best when they represent least, or least directly. The report of statistics adds value to the analysis inasmuch as it helps disclose causality. At least, one is supposed to be able to infer causality from certain statistical operations; that is the faith. Science moves beyond the directly phenomenal when it divines lawfulness from mere appearance.

Science makes use of appearance in order to construct an account of essence. It uses the word "statistics" to accomplish this work. I refuse any imputation of causality simply because history is open; people can choose and live against tabled patterns. Science deconstructs on the shoals of desire. Its resort to figural expression enhances the science aura in order to suggest a lawful world. But nothing turns the page into fact except readings that actualize the accounts presented as invariable. Ideology tells the truth because it successfully mystifies. People relinquish material equality in return for the promise of an afterlife; thus religion was right all along. People enter labor contracts seeming to promise fair exchange; thus bourgeois economic theory tells the truth. Here, people reproduce their fate reported ontologically in the text of science; thus science discloses the truth of the world. Writing forecloses political options only when it enters the social relations of representation; as ideology, it makes good on its frozen account through its own rhetorical virtuosity. The correlations become causal when we enact them as ontologically inevitable.

Percentages prepare the way for causal work. Inherently numbered, percentages initiate the science aura only reinforced by the display of statistics later on. Another feature of tables reporting correlations is that in occupying space, they leave space unoccupied. While a friend of nature cer-

TABLE 3

OLS REGRESSION OF RESPONDENT'S EDUCATION ON FATHER'S CLASS, SES, AND EDUCATION: EMPLOYED PERSONS OF NONFARM ORIGIN, AGES 25–64

Father's Class, SES, and Education	Men	Women
All respondents		
(N = 12,093 men, 5,937 women):		
Capitalist	.32*	.46*
	(.04)	(.07)
Manager	.90*	.70*
	(.07)	(.11)
Other supervisory	.76*	.77*
	(.07)	(.10)
Petty bourgeois	.09*	.11*
	(.04)	(.05)
Father's education	.51*	.52*
	(.01)	(.01)
R^2	.25	.24
Father's class = capitalist		
(N = 1,267 men, 570 women):		
Father's education	.40*	.32*
	(.03)	(.05)
Father's SES	.01*	.02*
	(.00)	(.00)
R^2	.18	.15
Father's class = manager		
(N = 519 men, 212 women):		
Father's education	.28*	.24*
	(.04)	(.06)
Father's SES	.06	[a]
	(.05)	
R^2	.09	.06
Father's class = other supervisory		
(N = 403 men, 229 women):		
Father's education	.25*	.30*
	(.05)	(.07)
Father's SES	.02*	.02*
	(.01)	(.01)
R^2	.13	.14
Father's class = petty bourgeois		
(N = 1,905 men, 954 women):		
Father's education	.48*	.35*
	(.03)	(.04)
Father's SES	.01*	.01*
	(.00)	(.00)
R^2	.25	.14
Father's class = nonsupervisory worker		
(N = 7,999 men, 3,972 women):		
Father's education	.38*	.42*
	(.02)	(.02)
Father's SES	.02*	.02*
	(.00)	(.00)
R^2	.23	.29

NOTE.—Column entries are unstandardized regression coefficients. Standard errors are in parentheses. All R^2's significant at $P < .001$.

[a] There is no variation in father's SES within this class origin.

* $P < .05$, one-tailed test.

FIGURE 7.7
Table Composed of Statistics

Source: See figure 7.5. 265. Used with Permission.

tainly in favor of saving trees from the sawmill, I note the failure to use space as an indulgence of figure's extravagant presence. Of course, statisticians would respond that certain tables (like the one in figure 7.8) dictate a given figuring of space. Some of the table must go unoccupied by number. But this only describes; it does not explain. Whatever the logic of presentation may be, the fact remains that journal science presents itself in a profligate way when it allows figure to dominate the space so thoroughly that territory is unclaimed by sign. Perhaps trivially disturbing from an ecological point of view, it is more telling as a mark of figure's hegemony in the text of science. It can afford to leave space blank simply because the presence of figure is so dominant. I doubt that journal editors scheme with compositional experts to achieve this effect; they do not view their work as aesthetic and political. Instead, they view the display of tables merely as communication. It is that, but it is also something else. Space has symbolic value when it is available to certain accounts but not others. Every journal editor worth her or his salt wants to economize in other sorts of ways; many journals stipulate that submitted manuscripts be limited to twenty or twenty-five typed pages. If editors must economize, they suggest cuts in prose, especially the literature review.

This is not to say that everything can be figure, at least today. Mediation is still required this side of the total transparency of representation. Hierarchy still requires differentiation against which to exist in counterpoint; for every superordinate, there must be a subordinate (or many). The logic of presentation is the crucial issue; figure displaces prose as evidence of intellectual advance. The new replaces and subordinates the old—figure over prose. One day, science promises, its work will be purely mathematical, ever the ideal of knowledge since the Enlightenment, especially, in social science aspiring to be "social arithmetic" (Condorcet). In the meantime, the shifting balance between prose and figure augurs the further scientization of local disciplines.

Figure 7.9 is another example of luxuriant figure. Even though prose counts for more than number in this table, number predominates, figuring certain causal paths among variables. What is striking is how laconically the numbers display themselves; there are only a few listed, leaving ample blank space. In disclosing little about the data, the table tells a lot; half a page of journal science is spent doing very little. Surely a more economic way could have been found to display or describe the same results. But tabular work is so privileged that space can be taken doing this. Imagine an authorial digression in the midst of the prose text—a qualification, clarification, cross-reference. Surely the editor would banish this to a footnote or excise it entirely. But tables can, in effect, ramble circuitously, commanding our attention even when they do little literary work.

Table 2. Sequential Logistic Regression Models Predicting Transition to Intercourse Between Rounds 1 and 2: White Females ($N = 280$)

Variables	Zero-Order Effects	Net Effects																
		1	2	3	4	5	6	7	8	9	10	11	12	13	14	15	16	17
Pubertal development	.48**	.21	.21	.22	.25	.28	.30	.30	.16	.15	.11	.12	.12	.12	.11	.08	.06	
Age	.73**	.65**	.63**	.59**	.50**	.49**	.50**	.37	.41*	.41*	.42*	.39	.39	.40	.45*	.46*	.49*	.54**
Attractiveness	.24*		.13	.13	.13	.14	.14	.04	−.00	−.00	−.02	−.03	−.03	−.03	−.04	−.04	−.04	
Parental household structure	−.70*			−.60*	−.73*	−.70*	−.58	−.61	−.58	−.56	−.53	−.50*	−.50	−.50	−.56	−.60	−.61	−.62
Same-sex friend's intercourse behavior	1.86**				1.71**	1.66**	1.54**	1.48**	1.23*	1.23*	1.25*	1.20*	1.19*	1.19*	1.30*	1.40*	1.33*	1.14*
Opposite-sex friend's intercourse behavior	1.30**					1.30**	1.33**	1.32**	1.18**	1.20**	1.19**	1.24**	1.24**	1.24**	1.27**	1.28**	1.28**	1.22**
School performance	−.65**						−.60*	−.56*	−.74**	−.74**	−.75**	−.82**	−.82**	−.81**	−.91**	−1.03**	−.96**	−.78*
Deviant behavior	.89**							.74**	.60**	.59**	.57**	.52*	.52*	.53*	.54**	.51*	.53*	.58**
Noncoital sexual experience	.94**								.69**	.68**	.71**	.69**	.69**	.66**	.63**	.59**	.50*	.42
Locus of control	−.21*									−.06	−.06	−.03	−.03	.03	−.01	.02	.02	
Future orientation	−.25*										−.15	−.14	−.14	−.13	−.13	−.17	−.16	
General sexual permisiveness	.42**											.30*	.30*	.30*	.29*	.22	.14	
Premarital sexual permissiveness	.10**												.00	.00	.00	.03	−.04	
Thinking about sex	.27**													.04	−.05	−.13	−.18	
"Turn-on" index	.16**														.14	.15	.16	.13
Subjective expected utility of sex	.02**															.01	.01	
Intention to have sex	.38**																.19	.24**
χ^2		24.03	24.92	28.80	41.99	60.11	64.83	81.69	93.94	94.11	95.36	101.39	101.39	101.48	104.66	108.34	111.93	106.79
d.f.		2	3	4	5	6	7	8	9	10	11	12	13	14	15	16	17	9

Note: Coefficients represent the change in the *log-odds* of making the intercourse transition per unit's increase in the independent variable.

FIGURE 7.8

The Report of Regression Coefficients in Table

From J. Richard Udry and John O.G. Billy, "Initiation of Coitus in Early Adolescence," *American Sociological Review* 52 (December 1987): 847. Used with permission.

376 AMERICAN SOCIOLOGICAL REVIEW

Table 1. Measurement Model of Educational Self-Direction

A. *Paths from First-Order Concepts to Indicators*

Indicator	Concept: Substantive Complexity of Schoolwork	Concept: Closeness of Supervision by Teachers
Complexity of coursework	.47*	—
Length of a "complete job"	.45*	—
Complexity of outside project	.40*	—
Complexity of extracurricular activities	.31*	—
Teachers tell student what to do	—	.23(t = 1.81)
Freedom to disagree with teachers	—	−.55*
Importance of doing exactly what teacher says	—	.26*
Freedom to choose outside-of-class schoolwork	—	−.41*

Correlation between concepts = −.43
χ^2 = 17.31 (for first-order model)
d.f. = 19
ratio = 0.91

B. *Paths to First-Order Concepts from the Second-Order Concept Educational Self-Direction*[1]

First-Order Concept	Educational Self-direction
Substantive complexity of schoolwork	.97*
Closeness of supervision by teachers	−.44

χ^2 = 11.43 (for "second-order" model)
d.f. = 8
ratio = 1.43

[1] This "second-order" measurement model must be embedded in a larger model to be identified; see text.
N = 189.
* p ≤ .05.

> assembly line a complete job may be to tighten two or three bolts; to an auto mechanic a complete job is to repair a car; to a coal miner, a complete job may be to load 18 tons. What do you ordinarily think of as a complete job in doing your schoolwork? How long does it take you to do a complete job?

We used duration of a complete job as an index of complexity, on the rationale that work that is readily divisible into short, discrete jobs tends to be less complex, while a job that takes a long time to complete is likely to entail a set of interrelated tasks and thus to be more complex.

3. *Complexity of an outside-of-class project.* Students were asked about the last term paper or other outside-of-class project they did that involved thinking and writing about something other than what was in their textbooks. From these descriptions, the complexity of the project was rated on a four-point scale: The student (1) simply brought some pre-existing exhibit or phenomenon to class; (2) turned to one or two encyclopedias, textbooks, or magazine articles; and the description of the project gives no evidence that the student did more than summarize or paraphrase this information; (3) used a variety of sources of a non-professional nature (e.g., magazines, encyclopedias, general books, texts), and the description provides some evidence that the student collated and organized these materials; or (4) used a variety of sources, including professional, technical, or other primary source material and/or gathered own data; and the description provides some evidence that the student evaluated and/or analyzed these materials to come to some independent conclusion.

4. *Complexity of extracurricular activities.* Students were asked about their school-related extracurricular organizational and athletic activities. School clubs and teams were rated on a three-point scale according to the substantive complexity of their principal activity, from least complex (generally non-academic activities, such as social clubs) to most complex (generally academically oriented activities, such as language clubs, and the preparation of

FIGURE 7.9

The Relationship Among Number, Prose and Empty Space

From Karen A. Miller, Melvin L. Kohn, and Carmi Schooler, "Educational Self-Direction and Personality," *American Sociological Review* (June 1986): 376. Used with permission.

This is not to propose simple solutions in order to save space and thus forests. That is not the point of my critique. Simply reducing the table to smaller dimensions, narrowing the white space, perhaps relegating it to an appendix is unimaginable within the larger literary context of journal science. Everything would have to change to make that possible. Science would have to lose its figural aura as well as its disciplinary sanction. It would have to intend all of its gestures and words poetically, rhetorically, politically, unashamed to be read metaphysically as well as empirically. Mere editorial policy will not undo the science aura, privileging figure over prose and thus diminishing thought. The whole fabric of the text would have to change, and with it (and thus through it) the world itself. At issue is positivist culture—ideology as material practice. While this culture is composed of infintesimally small gestures, such as the use of tabled data, it has numerous other features as well—books, television, religion, education. The totality must be changed for everything in it to change; ironically, though, it will not change unless the particulars change simultaneously. Strategy always confronts us with this irony. Piecemeal changes are just that.

Figure 7.10 extends this luxurious logic of tabular presentation. Open space abounds like an uncharted frontier. Here three tables occupying a whole page are separated by wide swaths of unprinted space. The sign-space ratio continues to fall. Science flexes its muscles and celebrates its stature by consuming space. This suggests a textual principle: Science's status is reflected both in the diminution of prose and of space used. As in mercantile and market capitalism, space is a source of value; the more of it held, the better. The text is governed by its own political economy; opponents skirmish for control of the terrain and thus value—here, the printed page.

As in capitalism, control brings both material and symbolic benefits; indeed, the more symbolic benefits are derived, the more material benefits accrue. Here, figure valorizes itself by reducing the prose component. As such, it tells the world its own way, figuring a figured world and thus making it more so. That leads me to another observation about the tables in figure 7.10. Prose virtually disappears from the tables. Only the titles above the tables contain words. The tables are composed of acronyms and numbers; prose decenters into technical code. Here we have the first pure example of figure claiming direct access to the world, unadorned by slow prose. Words disappear from the text; they are superfluous. Number and acronym take over.

Science wastes space in order to reduce prose, the remant of authorial thoughtfulness. Literary political economy governs the relationship between figure and prose. Both are text; figure, seeming marginal, is in fact the main text. Prose exists on the margin, if at all.

Table 2. Pooled GLS Estimates of Reform Effects on Prison and Jail Expansion (Standard Errors in Parentheses, $N = 192$)

PRIS =	31.79***	+.59	LPRIS***	−3.64	PROB	−12.42	PAR**	+.01	IND	+45.04	POP***
	(5.50)	(.05)		(5.06)		(4.78)		(5.45)		(3.57)	
JAIL =	6.17**	+.78	LJAIL***	+3.11	PROB	+.53	PAR	−7.70	IND*	+10.48	POP***
	(2.34)	(.04)		(2.55)		(2.58)		(3.01)		(1.57)	

* $p<.05$.
** $p<.01$.
*** $p<.001$.

Table 3. Pooled GLS Estimates of Urbanization, Literacy, and Reform Effects on Prison and Jail Expansion (Standard Errors in Parentheses, $N = 192$)

PRIS =	27.90***	+.73	LPRIS***	−6.65	URB	−77.97	LIT***	+3.26	PROB	−7.68	PAR	+4.39	IND	+75.31	POP***
	(5.92)	(.05)		(14.03)		(14.80)		(5.32)		(4.95)		(5.57)		(7.09)	
JAIL =	11.64***	+.69	LJAIL***	+32.25	URB***	+2.22	LIT	−2.62	PROB	−.94	PAR	−6.23	IND*	+6.80	POP*
	(2.50)	(.05)		(8.98)		(7.10)		(2.88)		(2.65)		(3.11)		(3.25)	

* $p<.05$.
** $p<.01$.
*** $p<.001$.

Table 4. Pooled GLS Estimates of Immigration, Wage-Labor, and Reform Effects on Prison and Jail Expansion (Standard Errors in Parentheses, $N = 192$)

PRIS =	33.30***	+.58	LPRIS***	−10.38	IMM	−125.00	WAGE*	+1.32	PROB	−7.05	PAR	−5.03	IND	+52.87	POP***
	(6.47)	(.05)		(28.49)		(57.22)		(5.48)		(5.04)		(5.57)		(4.27)	
JAIL =	11.51***	+.64	LJAIL***	+7.35	IMM	+107.78	WAGE***	−.98	PROB	−.09	PAR	−7.20	IND*	+6.64	POP***
	(2.97)	(.05)		(13.43)		(25.66)		(2.59)		(2.44)		(2.87)		(1.55)	

* $p<.05$.
** $p<.01$.
*** $p<.001$.

FIGURE 7.10

Empty Space as a Signifier of Number's Hegemony

From John R. Sutton, "Doing Time: Dynamics of Imprisonment in the Reformist State," *American Sociological Review* 52 (October 1987): 620. Used with permission.

Equation

Whereas tables flourish like mushrooms in journal science, the use of equations and formal models indicates an even higher stage of science. Indeed, one can plot exactly the point at which tables become formal models, raising science to the next higher plateau. Whereas tables can occasionally consist entirely of prose, they typically combine a few words with ample use of number. Some tables eliminate prose altogether in favor of number. Yet a particular kind of table even eliminates number—the ultimate act of semiotic subversion.

In the table in figure 7.11, number virtually disappears. Words are long gone, save for an incomprehensible title. The body of the table contains Greek letters, the semiotic currency of equations and formal models. A formal model postulates a state of being or a relationship without recourse to empirical data. Whereas the numbers in tables typically summarize data or correlations between and among variables, the Greek symbolism of formal models conveys the postulated states or relations of variables without any reference to the empirical world. Tabled data usually refer to the world, even if only indirectly. Tabled symbols bear no relation to reality whatsoever; they exist as abstract theoretical propositions, as such untested. Of course, pure mathematics is called *pure* for a reason. Its intellectual work need not be tested against an existing world of space and time. No one would disqualify mathematics from the pantheon of knowledges.

Yet the table in figure 7.11 is a literary artifact as well as a summary of formal mathematical reasoning. Its obtrusion on the page is a text in its own right. At least number promises to reveal something about the world—percentages, correlations between variables, income. The use of symbolism in formal mathematical reasoning only postulates things and relationships; they still must be tested against data to have more adequate representational meaning. Yet this table is read to be the epitome of science; obscure symbols are tabled in a patterned way, suggesting the esoteric revelation of a tabled, patterned world. Number and symbol are both figural gestures; as such, they exist on a semiotic continuum. Frequently, they belong together in journal science as in the frequent use of equations in the section on model specification.

In figure 7.12, equations consist of Greek symbols, letters, and numbers. They embody postulated relationships, to be checked in the model estimation section later on. As well, the equations interrupt prose. While formal models are occasionally tabled, as above, typically they interpose themselves between prose fragments, adding value to the science aura. Whereas the discipline of sociology makes frequent use of formal models captured in these unestimated equations, political science is even more dominated by formal modelers, especially on the part of people who write

Occupational Mobility

Structure and Quasi Symmetry

Observed flows are seldom symmetric. Indeed, a necessary condition for equation (3) to hold is that $F_{i.} = F_{.j}$ when $i = j$; that is, the expected marginal distributions of origins and destinations must be equal. Thus, if the expected marginal distributions are heterogeneous, as is typical in mobility tables, a strict exchange process cannot be operating; that is, symmetry cannot hold. Heterogeneous marginals do not rule out equal flows between some pairs of categories, a pattern known as "partial exchange" (Berger and Snell 1957). Equation (2) implies that $F_{ij} = F_{ji}$ for all i and j satisfying the condition that $\alpha_j = \alpha_i$ and $\gamma_{ij} = \gamma_{ji}$.

It is tempting to equate the asymmetric λ_{ij} parameters with structural mobility. By this definition, structural mobility is the residue from equal flows across the diagonal. This is also the definition used by Hutchinson (1958), which we have already rejected. Under our definition, the λ_{ij} parameters can be equated with structural mobility if and only if QS holds, that is, if and only if the lack of symmetry is due to inhomogeneous marginal effects but not due to asymmetrical association parameters. However, there are conditions analogous to the conditions of "partial exchange" when QS does not hold but some of the unreciprocated mobility in the table can be attributed to a subset of the α_j parameters and, thereby, to structural mobility.

We begin with equation (2), which expresses the λ_{ij} parameters in terms of marginal column effects (α_j) and asymmetric association parameters (γ_{ij}). QS arises if there is no asymmetric association, that is, if $\gamma_{ij} = 1$ for all i and j. The model has $(R - 1)(R - 2)/2$ *df*, and the usual fitting and testing methods (e.g., Bishop, Fienberg, and Holland 1975, pp. 286–91) apply. Table 1 presents an illustration of this parameterization for a 5×5 mobility table.

Under QS, the marginal shift between origins and destinations is repre-

TABLE 1

PARAMETERIZATION OF QUASI SYMMETRY WITH STRUCTURE AND EXCHANGE EFFECTS FOR A 5×5 TABLE

	DESTINATION				
ORIGIN	1	2	3	4	5
1	$\alpha_1\beta_1^2$	$\alpha_2\beta_1\beta_2\delta_{12}$	$\alpha_3\beta_1\beta_3\delta_{13}$	$\alpha_4\beta_1\beta_4\delta_{14}$	$\alpha_5\beta_1\beta_5\delta_{15}$
2	$\alpha_1\beta_1\beta_2\delta_{12}$	$\alpha_2\beta_2^2$	$\alpha_3\beta_2\beta_3\delta_{23}$	$\alpha_4\beta_2\beta_4\delta_{24}$	$\alpha_5\beta_2\beta_5\delta_{25}$
3	$\alpha_1\beta_1\beta_3\delta_{13}$	$\alpha_2\beta_2\beta_3\delta_{23}$	$\alpha_3\beta_3^2$	$\alpha_4\beta_3\beta_4\delta_{34}$	$\alpha_5\beta_3\beta_5\delta_{35}$
4	$\alpha_1\beta_1\beta_4\delta_{14}$	$\alpha_2\beta_2\beta_4\delta_{24}$	$\alpha_3\beta_3\beta_4\delta_{34}$	$\alpha_4\beta_4^2$	$\alpha_5\beta_4\beta_5\delta_{45}$
5	$\alpha_1\beta_1\beta_5\delta_{15}$	$\alpha_2\beta_2\beta_5\delta_{25}$	$\alpha_3\beta_3\beta_5\delta_{35}$	$\alpha_4\beta_4\beta_5\delta_{45}$	$\alpha_5\beta_5^2$

363

FIGURE 7.11
Table Composed of Greek Letters as Signifiers

From Michael E. Sobel, Michael Hout, and Otis Dudley Duncan, "Exchange, Structure, and Symmetry in Occupational Mobility," *American Journal of Sociology* 91, no. 2 (1985): 363. Used with permission.

WORKER ATTACHMENT AND WORKPLACE AUTHORITY 641

that a worker will voluntarily dissolve his contract with his current employer. Compared to other candidates, this definition is consistent with the framework developed above, affectively neutral, conceptually homogeneous, and conveniently measured by common verbal or behavioral indicators of quit propensities. Furthermore, it pits the authority-cost hypothesis directly against the CV approach, since the latter focuses almost exclusively on job quitting. Of course, quitting is also the major behavioral outcome studied within the JSOC approach, so common ground is established here as well.

Model Specification

In both data sets attachment is indicated by ordinal measures of quit propensities. Experiments were conducted with various codings (linear, binary) and statistical models (linear regression, linear probability and binary logistic regress); all yielded qualitatively similar results. The statistical model settled on treats the observed ordinal responses as discrete categorical expressions of an underlying continuous distribution of quit propensities. Let quit propensities be determined by

$$Q_i = \beta' X_i + e_i,$$

where X_i is a set of explanatory capital-value and authority-cost variables, β is a vector of parameters, and the e_i are disturbances. Let the observed ordinal variable Q^*_j, $j = 1,2, \ldots m$, have values determined by

$$Q^* = j \text{ if } \alpha_{j-1} < Q < \alpha_j$$

where the α_j are constants, with $\alpha_0 = -\infty$, $\alpha_m = +\infty$, and $\alpha_0 < \alpha_1 < \ldots < \alpha_m$. Then the probability distribution of Q^* is given by,

$$P(Q^* = j) = F(\alpha_j - \beta' X_i) - F(\alpha_{j-1} - \beta' X_i)$$

where F is the cumulative distribution function of the e_i. If F is the cumulative normal, then this is the so-called ordered probit model (McKelvey and Zavoina, 1975; Winship and Mare, 1984). Following Cox (1970, pp. 103–104) and McCullagh (1980), assume instead that F is the logistic cumulative distribution function, so that the probability model of Q^* as a function of the explanatory variables is

$$P(Q^* = j) = \frac{exp(\alpha_j - \beta' X_i)}{1 + exp(\alpha_j - \beta' X_i)} - \frac{exp(\alpha_{j-1} - \beta' X_i)}{1 + exp(\alpha_{j-1} - \beta' X_i)}$$

Estimation of the coefficient vector β and the cutpoints α_j proceeds within a maximum likelihood framework. Tests of the authority-cost hypothesis take $Q = q(X_k)$ as the null model and $Q = q(X_k, X_c)$ as the alternative, where the explanatory variables are partitioned into their capital-value and authority-cost components.

ANALYSIS OF FIRM SAMPLE

The data pertain to male "management" personnel of what at the time (1960) was the largest single enterprise of a major public utility holding company in the U.S. A 75 percent response rate to questionnaires distributed to all personnel yielded a sample size of 1091 for all analyses.

The dependent variable indicates the propensity to accept an offer from another employer. It is based on the question: *Suppose you were offered a job exactly like your present position with another company located nearby but with a 20 percent increase in salary. Would you accept it?* with responses on a four-point scale ("definitely" to "definitely not"). This measure has several advantages in terms of internal validity; chief among them is that it eliminates a weakness of virtually all quit research (Viscusi, 1980; Parsons, 1972; Freeman, 1980) by controlling directly for variation among workers in their knowledge of the distribution of alternative contracts and their opportunity for forming contracts with other employers.[15]

As with most CV research, the principal indicators of the capital value of the current contract are linear and quadratic terms for salary and years of service with the firm. Age measures the length of a worker's remaining work life, and is used as an inverse measure of the expected long-run gains from quitting (Viscusi, 1980; Burdett, 1978). The total number of positions held in the present firm (POSITIONS1) and the number of positions held previously in other firms (POSITIONS2) are used as indicators of a worker's location in his career trajectory. Note that the firm-specific nature of the sample and the dependent variable itself automatically control for many sources of variation in capital-value factors that economic research on quitting has controlled statistically (e.g., region, industry, unemployment rate, alternative opportunities, etc.).[16]

The authority-cost variables are all based on worker perceptions of governance practices as being in conformity with various aspects of legal-rational rule. The first, based on Selznick's

[15] This item may also have a high degree of criterion validity in the sense that it would be highly correlated with responses to an actual offer. I base this judgement on the fact the usual intention-to-quit items used by organizational psychologists tend to be highly correlated with actual turnover (Miller et al., 1979; Mobley et al., 1978; David, 1974; and Steel and Ovalle, 1984). The item also has a high degree of construct validity insofar as it behaves with respect to capital value variables just like measures of actual quit behavior used in economic research.

[16] Preliminary analyses included measures of schooling and domestic responsibilities (marital status and number of kids), but these variables were statistically inconsequential.

FIGURE 7.12
Equations as "Text" on the Journal Page

From Charles N. Halaby, "Worker Attachment and Workplace Authority," *American Sociological Review* 51 (October 1986): 641. Used with permission.

in game theory. A perusal of leading American political science journals, such as *American Political Science Review* and, to a lesser extent, *American Journal of Politics,* reveals the fanatical use of formal models. They appear everywhere. Only sometimes are they supported by data—numbered. Contemporary political science, even more than sociology in this sense, seems to have become a mathematics discipline. Intellectual work becomes the adumbration of proofs; validity reposes in what mathematicians call elegance.

Equations litter the text, especially in the model specification section. Sometimes they overwhelm it. Figure 7.13 presents a series of differential equations, concluding the model specification section of an article. One wonders how essential they are for the train of the argument. Would words not suffice to summarize the hypotheses to be tested? By semiotizing hypothetical reasoning, journal science both devalues prose as a medium of reason and redoubles the norm of figural expression. Differential equations are science; science requires differential equations. No matter that few can understand them sufficiently to work with them. They appear in the text for their own sake. They need no elaboration. Tellingly, the equations "quoted" end a section of the article. Words are unnecessary to explain them further. Formal mathematical reasoning is sui generis; only prose needs to be explicated, translated, semiotized.

In principle, words can be translated into figure, and figure can be rendered in prose. Both words and figure are semiotic codes, signs signifying only in relation to whole systems of meaning. Saussure distinguished between word and speech, suggesting that words gather meaning only in systems of discourse. Both words and figures are in effect words; as signs, they must be explicated in reference to other signs. Systems of meaning have no outside. Words can be transformed into other signs, and signs can be transformed into other words. Linguistic transformation is limitless in its possibilities. Yet the text figures words, but not the other way around. The differential equations end the section in figure 7.13; they do not begin it, requiring words to unravel them.

Prose points to its semiotization; once figured, science rests. One could imagine a mixed literary strategy whereby signs transform each other dialogically, prose enlightening figure and figure punctuating and thus clarifying prose. Nothing inherently privileges prose over other forms of literary expression. In a better world it would coexist with other representational and nonrepresentational discourses. But in science prose takes a back seat to figural work. While it may buttress it, it can expect no buttress of its own. The differential equations, not the prose preparing the way for them, are the epitome of journal science.

The page shown in figure 7.14 could easily be culled from a mathematics textbook. The sociology is virtually absent, save for the very top of the dynamics of areas that are urban, or Protestant, or both urban and Protestant, or to explore the nature of partisan competition in areas that have high concentrations of workers, farmers, or small business owners. In sum, it is desirable to condition the values of the parameters to reflect additional social information. This can be accomplished directly by writing each parameter in the form

$$a_0 + a_1X, \tag{10}$$

thus making each parameter a linear function of X (the social variable). (Note that in the case in which $X = 0$, the original model is recovered.) The social variable can be an interactive variable if more than one social characteristic conditions the model simultaneously, for example, the case of Protestant urban areas with high levels of petty bourgeoisie. The entire system now can be written as

$$\begin{aligned} dR/dt = {} & R[(f_0 + f_1X)\,C - (b_0 + b_1X)\,L \\ & + (m_0 + m_1X)\,N] \\ & + (v_0 + v_1X) && (11) \\ dC/dt = {} & C[(s_0 + s_1X)\,N - (f_0 + f_1X)\,R \\ & - (a_0 + a_1X)\,L] \\ & + (k_0 + k_1X) && (12) \\ dL/dt = {} & L[(a_0 + a_1X)\,C + (q_0 + q_1X)\,N \\ & + (b_0 + b_1X)\,R] \\ & + (j_0 + j_1X) && (13) \\ dN/dt = {} & (w_0 + w_1X) - N\,[(m_0 + m_1X)\,R \\ & + (q_0 + q_1X)\,L \\ & + (s_0 + s_1X)\,C]. && (14) \end{aligned}$$

ESTIMATING THE SYSTEM

The techniques for evaluating the stability of the system are well established (see May 1974). All such techniques require, of course, that all parameter values be obtained first. The estimation of the above parameters is a nontrivial problem that cannot be solved by regression techniques. First, there is no way to solve for R, C, L, or N explicitly. Thus, we are left with the problem of estimating four models that must remain in derivative form. Linearizing techniques for less complex models have been suggested by Coleman (1981) and Tuma and Hannah (1984). However, such techniques require the uncoupling of the equations and are pursued to recover known statistical properties of the estimators. Several useful techniques, often used to solve practical engineering problems, employ iterative approximations to definite integration.[4] The techniques, as they are used here, are described in the Appendix A.

The data in this analysis are for the approximately one thousand *Kreise* in all of Germany in the Weimar period.[5] *Kreise* are comparable to U.S. electoral districts, but census information (such as religion and occupation) is also collected at the level of the *Kreise* in Germany.

The voting variables are all measured as proportions of the population that support the rightists, centrists, leftists, or are nonvoters. Supporters for the smaller parties (invariably a small proportion of the total vote) are not included. The period examined spans from 1928 to July 1932, from the genesis of the Nazi electoral movement to the time before the Nazi seizure of power. Each period between elections is examined separately. Thus, we estimate the dynamics of the electoral system from May 1928 to September 1930, then again from September 1930 to July 1932.

The social data are also measured as proportions of the total population within each *Kreis*. However, the proportions are transformed by standardizing each of the social variables to have a zero mean and a standard deviation of one. This transformation encourages the intuitively appealing interpretation of these data as measures of the social "atmosphere," acknowledging the notion that small changes in proportions for particular populations can signal large changes in the informational biases of the social milieu. The social variables used are religion (proportion of the population that is Protestant or Catholic), urbanization (a population density measure is used: a low value signifies a rural condition, a high value, a more urban environment), and occupation. Previous analyses of the Weimar period have indicated that these social variables are valuable in characterizing many conditioning aspects of the Weimar political environment (Hamilton 1982; Brown 1982). The first occupational variable is the proportion of the population operating farms (thus estimating the size of the peasantry). The second occupational variable is a measure of the size of the petty bourgeoisie, specifically, the proportion of the population engaged in trade and transportation.

The dependent variables are the electoral strengths of each of the groups (rightist, centrist, leftist, and nonvoter) over time. The models are evaluated in terms of their ability to explain a change in voter support for, say, the rightists

[4] For a lucid discussion of such techniques, see Hamming (1971).

[5] The voting and religious data used here were provided by the Inter-University Consortium for Social and Political Research. Of course, this organization bears no responsibility for any errors of interpretation of these data that may have occurred here.

FIGURE 7.13

The Semiotization of Hypothetical Reasoning

From Courtney Brown, "Voter Mobilization and Party Competition in a Volatile Electorate," *American Sociological Review* 52 (February 1987): 64. Used with permission.

page, where the words *Democrat* and *Republican* appear. Although this page is taken from the article's appendix, it enjoys a kind of prominence. In case the article was not mathematized enough, the authors add a few appendix pages to clinch the issue. Prose joins these figural items, but it is essentially figured itself; little in the way of sociological interpretation or analysis is to be found. Of course, working scientists would say that the appendix is exactly the place for these methodological afterthoughts. Yet the body of the article is no less methodological in its interest. The appendix is merely the icing on the cake. The very last piece of text (see figure 7.15) before the references tables these Greek signs in a virtually artistic representation—but of just what?

Surely the authors do not intend representation in these peaks and valleys of Greek letters. Yet the table reeks of mathematical and figural elegance; something important is going on. This opens the next section of the chapter, where I consider graphic representation as a crucial component of the science aura. Again, figural usages are often heterogeneous in their literary work. Tables, equations, graphs, models, and pictures frequently combine as methodological quantity becomes methodological quality.

Social science writes this way in order to raise their disciplines to the rigorous level of the natural sciences and mathematics. But that assumes that the sciences and mathematics deserve to be there. This presupposes a standard of scientificity hinged on the figural displacement of prose; the less prose and the more figure and number, the higher science is judged to be. But one can read figure to be saying something else, as I am trying to do here. The displacement of words by figure in itself does not signify intellectual progress; indeed, it frequently signals regress when figure reduces the imaginative role of authorial subjectivity. The old-fashioned author could meander, circle, reflect, double back, and then interrogate these interrogations in the process of reaching understanding. Figure and number merely intend to reflect, to represent, a figured and numbered order. While the peaks and valleys do not suggest a world familiar to us, in having shape they suggest a shapefulness to the analysis in question.

Journal science accordingly attempts to give the world direct access to the text itself. The peaks and valleys composed of Greek letters suggests both a higher level of mathematical reasoning and a representationality supporting science's claim to make metaphysics present on the page. Using shape and form as literary devices, science suggests a shaped, formed, figured, numbered, and tabled world to which these particular literary devices afford direct access. Of course, science pretends that table, equation, graph, model, and picture are methodological techniques, not rhetorical gestures. That is precisely an aspect of its unconsciousness. The peaks and valleys intend significance, if not a one-to-one correspondence to particular peaks and valleys. Rather the table in figure 7.15 suggests an

CHURCH ATTENDANCE IN THE UNITED STATES 343

one party or the other, (2) weak Democrat or Republican, or (3) strong Democrat or Republican. The logit regression of loyalty and item difficulty (Y^P) on the log-odds on being in category k of variable P, relative to being in category $k-1$ is given by:

$$\Phi_{ki}{}^P = \alpha_k{}^P + \log(x_i) - \log(y^P). \quad (4)$$

Equations (3) and (4) are unobservable, but they have implications for the observed distributions of A and P. The first step in deriving those implications is to take the antilogs of both sides of both equations:

$$\exp(\Phi_{ki}{}^A) = \Omega_{ki}{}^A = \mu_k{}^A x_i/y^A \quad (3')$$

$$\exp(\Phi_{ki}{}^P) = \Omega_{ki}{}^P = \mu_k{}^P x_i/y^P, \quad (4')$$

where $\mu_k{}^A) = \exp(\alpha_k{}^A$ *and* $\mu_k{}^P = \exp(\alpha_k{}^P)$. Since we do not know the scale of X, we identify the parameters of (3′) and (4′) and simultaneously set the scale of X by setting $\mu_1{}^A = \mu_1{}^P = 1$. The sense in which setting the μ_1 parameters equal to one sets the scale of X is explained below.

Following Duncan (1985a, 1985b) we can express the conditional probability of subject i falling into category k of variable A or P as a function of X and the parameters of equations (3′) and (4′):

$$\text{prob}(A_i = 0|x_i) = 1/C_i \quad (5a)$$

$$\text{prob}(A_i = 1x_i) = x_i/C_i \quad (5b)$$

$$\text{prob}(A_i = 2|x_i) = \mu_2{}^A x_i{}^2/C_i \quad (5c)$$

$$\text{prob}(A_i = 3|x_i) = \mu_2{}^A \mu_3{}^A x_i{}^3/C_i \quad (5d)$$

$$\text{prob}(P_i = 0|x_i) = 1/D_i \quad (6a)$$

$$\text{prob}(P_i = 1|x_i) = yx_i/D_i \quad (6b)$$

$$\text{prob}(P_i = 2|x_i) = y^2\mu_2{}^P x_i{}^2/D_i \quad (6c)$$

$$\text{prob}(P_i = 0|x_i) = y^2\mu_3{}^P x_i{}^3/D_i, \quad (6d)$$

where $y = (y^A/y^P)$ and where $C_i = (1 + x_i + \mu_2{}^A x_i{}^2 + \mu_2{}^A\mu_3{}^A x_i{}^3)$ and $D_i = (1 + yx_i + y^2\mu_2{}^P x_i{}^2 + y^2\mu_2{}^P \mu x_3{}^P x_i{}^3)$ are adjustments that fix the sums of the conditional probabilities at one. These probabilities are undefined if $x_i = 0$. However they approach the limits:

$$\text{prob}(A_i = 0|x_i = 0) = 1$$

$$\text{prob}(A_i = 1|x_i = 0) = 0$$

$$\text{prob}(A_i = 2|x_i = 0) = 0$$

$$\text{prob}(A_i = 3|x_i = 0) = 0$$

$$\text{prob}(P_i = 0|x_i = 0) = 1$$

$$\text{prob}(P_i = 1|x_i = 0) = 0$$

$$\text{prob}(P_i = 2|x_i = 0) = 0$$

$$\text{prob}(P_i = 3|x_i = 0) = 0.$$

We are interested in the values of five parameters: y and the four μ parameters. As noted above, y measures the relative difficulties of A and P. An estimate of $y > 1$ indicates that A is more difficult than P; $y < 1$ indicates the converse. The interpretation of the μ parameters hinges on our earlier claim that setting $\mu_1{}^A = \mu_1{}^P = 1$ fixes the scale of X. To understand this claim, note that by substituting (5a) and (5b) into (1), we obtain: $\Omega_{1i}{}^P = \log(x_i)$. If subject i is indifferent between never attending church and attending yearly, that is, if the probability in (5a) equals the probability in (5b), then $\Phi_{1i}{}^P = 0$, implying that $\log(x_i) = 0$, so $x_i = 1$ when the probability of "never" equals the probability of "yearly." Since the probability of "never" is one and the probability of "yearly" is zero when $x_i = 0$, one unit on the X-scale is the amount of increase in loyalty required to raise the probability of yearly attendance and lower the probability of no attendance to the point of equality.

At that point where $X_i = 1$, the full set of odds for A is

$$\Omega_{1i}{}^A = 1$$

$$\Omega_{2i} = \mu_2{}^A$$

$$\Omega_{3i}{}^A = \mu_3{}^A,$$

so the $\mu_k{}^A$ parameters give the value of their respective odds at the point where $x_i = 1$. If $\mu_k{}^A > 1$, then the probability of attendance level k is greater than the probability of "never" and "yearly"; if $\mu_k{}^A < 1$, then the converse is true. For P,

$$\Omega_{1i}{}^P = y$$

$$\Omega_{2i}{}^P = y\mu_2{}^P$$

$$\Omega_{3i}{}^P = y\mu_3{}^P,$$

so y is the ratio of the probability of "leaning" to the probability of being "independent" when x_i is one. A little algebra reveals that $\Omega_{1i}{}^P = 1$ when $x_i = 1/y$. The interpretation of the $\mu_k{}^P$ is similar to the interpretation of the $\mu_k{}^A$.

The key assumption in deriving the implications of a Rasch model for an observable cross-tabulation is "local independence," that is, the assumption that the joint conditional probability that $A_i = k$ and $P_i = k'$ for a constant value x_i (denoted $p_{kk'i}$) is the product of the two constituent conditional probabilities:

$$p_{kk'i} = \text{prob}(A_i = k|x_i) \times \text{prob}(P_i = k'|x_i)$$

$$= (\pi_k \mu_k{}^A)(\pi_{k'} \mu_{k'}) y^{k'} x_{ikk'}/(C_i D_i). \quad (7)$$

The top panel of Table A1 contains expressions of this type for the 4 × 4 table at hand. To estimate the identified y and μ parameters, we need an expression of expected frequencies under the model in terms of the y and μ parameters. Expected frequencies are obtained by

FIGURE 7.14

Prose Replaced by the Semiotics of Formal Modeling

From Michael Hout and Andrew M. Greeley, "The Center Doesn't Hold: Church Attendance in the United States, 1940–1984," *American Sociological Review* 52 (June 1987): 343. Used with permission.

344 AMERICAN SOCIOLOGICAL REVIEW

summing joint conditional probabilities over subjects, that is, $F_{kk'} = \Sigma_i p_{kk'i}$, as shown in the second panel of Table A1. The $S_t(t = 0, \ldots, 6)$ parameters are called "score group" parameters. They contain all of the information about heterogeneity on the underlying loyalty dimension, that is, they are the only parameters that are a function of the individual scores (x_i). Note that all of the association between attendance and partisanship is due to heterogeneity on the underlying loyalty dimension, that is, all of the odds ratios for 2 × 2 subtables of expected frequencies are functions of the score group parameters. For example,

$$\theta_{23} = \log(F_{23}F_{34}/F_{24}F_{33})$$
$$= \log(S_3S_5/S_4^2).$$

If loyalty is constant, that is, if $x_i = x$ for $i = 1, \ldots, N$, then S_t is a constant for each t (= 0, ..., 6), and the nine odds ratios for 2 × 2 subtables formed from adjacent rows and columns are all one, that is, $\theta_{ij} = 1$ ($i = 0,1,2$; $j = 0,1,2$). On the other hand, all else being equal, an increase in the variance of X will increase the strength of the association between A and P. In that sense, then, the association between attendance and political partisanship is spurious when loyalty is controlled.

Table A1. Conditional Joint Probabilities and Expected Frequencies for a Latent Variable Model of the Effect of Loyalty on Church Attendance and Political Commitment for Catholics

Church	Political Commitment			
Attendance	0	1	2	3
Conditional Joint Probabilities				
0	$1/Z_i$	yx_i/Z_i	$\mu_2{}^P y^2 x_i{}^2/Z_i$	$\mu_2{}^P \mu_3 P y^3 x_i{}^3/Z_i$
1	x_i/Z_i	$yx_i{}^2/Z_i$	$\mu_2{}^P y^2 x_i{}^3/Z_i$	$\mu_2{}^P \mu_3{}^P y^3 x_i{}^4/Z_i$
2	$\mu_2{}^A x_i{}^2/Z_i$	$\mu_2{}^A yx_i{}^3/Z_i$	$\mu_2{}^P \mu_2{}^A y^2 x_i{}^4/Z_i$	$\mu_2{}^P \mu_3{}^P \mu_2{}^A y^3 x_i{}^5/Z_i$
3	$\mu_2{}^A \mu_3{}^A x_i{}^3/Z_i$	$\mu_2{}^A \mu_3{}^A yx_i{}^4/Z_i$	$\mu_2{}^P \mu_2{}^A \mu_3{}^A y^2 x_i{}^5/Z_i$	$\mu_2{}^P \mu_3{}^P \mu_2{}^A \mu_3{}^A y^3 x_i{}^6/Z_i$
Expected Frequencies				
0	S_o	yS_1	$\mu_2{}^P y^2 S_2$	$\mu_2{}^P \mu_3{}^P y^3 S_3$
1	S_1	yS_2	$\mu_2{}^P y^2 S_3$	$\mu_2{}^P \mu_3{}^P y^3 S_4$
2	$\mu_2{}^A S_2$	$\mu_2{}^A yS_3$	$\mu_2{}^A \mu_2{}^P y^2 S_4$	$\mu_2{}^A \mu_2{}^P \mu_3{}^P y^3 S_5$
3	$\mu_2{}^A \mu_3{}^A S_3$	$\mu_2{}^A \mu_3{}^A yS_4$	$\mu_2{}^A \mu_3{}^A \mu_2{}^P y^2 S_5$	$\mu_2{}^A \mu_3{}^A \mu_2{}^P \mu_3{}^P y^3 S_6$

REFERENCES

Altheizer, Thomas and William Hamilton. 1966. *Radical Theology and the Death of God*. Indianapolis: Bobbs-Merrill.

Alwin, Duane F. 1984. "Trends in Parental Socialization Values: Detroit, 1958–83." *American Journal of Sociology* 90:359–82.

______. 1986. "Religion and Parental Child-Rearing Orientations: Evidence of a Catholic-Protestant Convergence." *American Journal of Sociology* 92:412–40.

Baker, R.J. And J.A. Nelder. 1978. *The GLIM System: Release 3*. Oxford: Numerical Algorithms Group.

Berger, Peter. 1969. *A Rumor of Angels*. New York: Doubleday.

Bergesen, Albert and Mark Warr. 1979. "A Crisis in the Moral Order: The Effects of Watergate upon Confidence in Social Institutions." Pp. 277–95 in *The Religious Dimension: New Directions in Quantitative Research*, edited by Robert Wuthnow. New York: Academic Press.

Caplow, Theodore, Howard M. Bahr, and Bruce A. Chadwick. 1983. *All Faithful People: Change and Continuity in Middletown's Religion*. Minneapolis: University of Minnesota Press.

Clogg, Clifford, C. 1978. "Adjustment of Rates Using Multiplicative Models." *Demography* 15:523–40.

Converse, Philip E. 1976. *The Dynamics of Party Support: Cohort Analyzing Party Identification*. Beverly Hills, CA: Sage.

Cox, Harvey. 1965. *The Secular City*. New York: Macmillan.

Davis, James A. and Tom Smith. 1984. *General Social Survey Cumulative Codebook: 1972–84*. Storrs, CT: Roper Center.

Duncan, Otis Dudley. 1985a. "Measurement and Structure: Strategies for Design and Analysis of Subjective Survey Data." Pp. 179–230 in *Surveying Subjective Phenomena*, vol. 1, edited by Charles F. Turner and Elizabeth Martin. New York: Russell Sage Foundation.

______. 1985b. "Rasch Measurement: Further Examples and Discussion." Pp. 367–403 in *Surveying Subjective Phenomena*, vol. 2, edited by Charles F. Turner and Elizabeth Martin. New York: Russell Sage Foundation.

Fienberg, Stephen E. 1980. *The Analysis of Cross-Classified Categorical Data*. 2nd ed. Cambridge, MA: MIT Pess.

Fienberg, Stephen E. and William M. Mason. 1978. "Identification and Estimation of Age-Period-Cohort Effects in the Analysis of Disrete Archival Data." Pp. 1–63 in *Sociological Methodology* 1979, edited by Karl F. Schuessler. San Francisco: Jossey-Bass.

Gallup, George H. 1972. *The Gallup Poll: Public Opinion 1935–71*. New York: Random House.

Gallup, George, Jr. 1985. *Religion in America: Fifty Years, 1935–1985*. Princeton: American Institute of Public Opinion (Gallup Report #236).

Gilkey, Langdon. 1969. *Naming the Whirlwind*. Indianapolis: Bobbs-Merrill.

Glock, Charles Y. and Rodney Stark. 1965. *Religion and Society in Tension*. Chicago: Rand-McNally.

Goodman, Leo A. 1970. "The Multivariate Analysis of Qualitative Data: Interactions Among Multiple Classifications." *Journal of the American Statistical Association* 65:226–56.

______. 1972. "A General Model for the Analysis of Surveys." *American Journal of Sociology* 77:1035–86.

______. 1981. "Three Elementary Views of Log-Linear

FIGURE 7.15
Formal Modeling as a Mode of Representation

Source: See figure 7.14. 344. Used with Permission.

essential formfulness—peakness and valleyness—to the world, authorizing its disclosure of an inert world.

Methodology makes way for metaphysics—ontological presence—when it employs rhetorical devices resembling art more than science. This is not to say that art is merely fancy, for art tells its own truth.[8] The only literary difference between art and science is that science claims validity methodically whereas art presupposes it. This is a trivial difference; it allows us to locate many literary products heretofore assumed to be different in kind along the same epistemological continuum. Both *A Tale of Two Cities* and *Capital*—no less than journal science—disclose a world; whether they do it accurately is another question. In every instance, science takes literary license in putting pen to paper. The results must always be crafted; they are not sui generis. So in saying that methodology becomes metaphysical, I am not denigrating art as a form of talk intending validity. I am only impugning metaphysics as an arbitrarily ahistorical version of the world. Art and science join forces against metaphysics not on grounds of its speculativeness, for every version essays a world. They oppose methodology's aversion to history and thus the possibility of an open future. Methodology is the art (and thus politics) of science that wants to deceive us about its inherently literary nature. Instead, the figural doodles of science pretend an access to the truth denied to art and politics. This is a conceit on the part of methodology, especially when methodology can be read against itself as a gestural—hence political—text.

The table reproduced in figure 7.15 intends itself to be read as a refraction of the world through a semiotic discourse ensuring an undistorted account. The seeming representationality of the peaks and valleys duplicates not the world as such but its essence adduced from the methodologically driven analysis of data. No one believes that the peaks and valleys reflect the Grand Canyon per se; they are led to believe that they represent ontological presence of one sort or another. At least, the figural representations claim access to the world denied to the passionate, pouting, political zealotry of prose. As such, they are counterposed to what science fears most in a positivist age—polemic. In an age allegedly beyond ideology, that is the deepest ideology of all; in fact, differences proliferate and should be acknowledged. Otherwise the representation of homogeneity—the sine qua non of metaphysics—guarantees a metaphysical outcome. The text read as a distortionless and distinterested account distortingly and interestedly advocates a world it pretends only to reflect.

Graph

The table in figure 7.15 leads journal science into a whole realm of gestures including graphs, models, and pictures. Beyond tabular representation (except of the sort just displayed), science suggests the contours of the world itself. Even numbered data are thought to mediate; here, tables take on a second dimension as they become forms themselves. As such, like the table in figure 7.15 they appear cleanly to reflect a metaphysical world order. If they do not capture the things themselves—mountains, flux, flow—they promise thingness: mountainness, fluxness, flowness. Of course, the linear thinker doubles over with hilarity at words like these, suspecting the account of a frivilous disrespect for language. But the figures in journal science *are texts;* they are meant to be read as deauthored accounts of an ontological world, thus boosting the article's science aura. It is an old-fashioned premise that only the words count in science; much more is going on in the text than that. Indeed, the words are increasingly overwhelmed with the figural embellishments taking over the role of epistemological constitution. What is text must be decided empirically—here, with reference to the constitutional work being done. If words take a back seat to figure, that is a telling text in its own right. In any case, literary analysis cannot afford to assume that journal science proceeds like a story. Narrative has become methodology—and methodology figure.

The graphs in figure 7.16 typify graphic work found throughout journal science. Graphic work is not mathematically more advanced than the report of certain numbers in tables; it is, however, a higher order of figure, suggesting if not the world then worldness. For the most part, tables organize the special case of figural work dominated by the display of quantity.

The graph intends a two-dimensional world, adding dimension to the mediateness of the tabular report of number. While number is inherently high science, it is not necessarily the highest form, especially when it is supplemented by graphed, modeled, and pictured work. The graphs in figure 7.16 look *like* something, thus suggesting the text's representationality—and therefore guaranteeing its scientificity. The graphs do not have actually to be something; they must only suggest a two-dimensional world, to which the reader gains access through the scientific text. Of course, science would say otherwise, arguing that graphic display only enhances "communication," the same argument used to defend tabular representation. Indeed, the ability to "read" graphs, tables, charts, and pictures is basic to our civilization's concept of intelligence; it represents either analytical or analogical ability. Any high school student knows that this intelligence is highly valued. Intelligence and aptitude tests are littered with questions about the implications of these sorts of figures.

I do not have to demonstrate that figures such as these can be read as texts. We already know that from educational testing. Journal science wants to suppress the constitutionality of figure in order for figure to be even more constitutional than it is on the surface. Whereas we believe prose is in the saddle, it is even more disempowered by the gestural work going on

AGE STRUCTURE AND CRIME 173

of U.S. arrests for motor vehicle theft have shown substantial invariance in form during the past four decades.

A somewhat different conclusion emerges from Figure 2, which gives the five-year age distributions of persons arrested per 100,000 population in the United States for murder and nonnegligent manslaughter in 1960 and 1980.[3] Specifically, whereas two age groups, namely, ages 20–24 and 25–29 have higher rates than those of any other age groups in 1960, they are joined by the 15–19 age group in 1980. Again, these characteristics also hold for age curves from other years. The age curve of arrests for 1950 is similar to that for 1960, and the curve for 1970 is similar to that for 1980. In brief, age curves of arrests for homicide changed in the 1960s in such a way as to raise the rates for 15–19 year-olds to levels close to those of persons in their twenties, indicating the possibility of a cohort effect, as suggested by Greenberg (1985).

Consistent with the findings of Hirschi and Gottfredson (1983), the rates of the peak age categories in Figure 2 are not as dramatically higher than those of other age categories as is the case in Figure 1 and the homicide arrest rates taper off more slowly with increasing age than do the vehicle theft rates. These differentials also are found for other property and violent offenses during the past four decades. In order to maximize the validity and accuracy of the temporal analyses described below, however, we shall limit our discussion primarily to vehicle theft and homicide.[4]

One implication of the unimodal skewed age curves in Figures 1 and 2 is that an increasing percentage of the population in the teen and young adult age categories should produce increasing crime rates. This proposition was verified by a number of studies (Sagi and Wellford 1968; Ferdinand 1970; Wellford 1973) that used demographic standardization techniques (on age-specific arrest rates) to conclude that 40 to 50 percent of the rise in the total FBI Index crime rate during the 1960s could be accounted for by changes in age composition.

To illustrate this relationship, Figures 3 and 4 show post–World War II trends in age composition and crime rates. Figure 3 exhibits annual vehicle theft rates per 100,000 population

[3] Data for Figure 2 were obtained from the same sources and processed in the same way as those for Figure 1 (see footnote 2).

[4] Cantor and Cohen (1980) found that UCR data on homicide rates are relatively unaffected by sampling and coverage problems from 1946 to recent years. In addition, victimization surveys consistently find that motor vehicle theft is the most completely reported property crime (Skogan 1976). Given that motor vehicles are relatively expensive pieces of property and their theft must be reported for victims to collect insurance settlements, this finding is intuitively meaningful. In brief, we feel that the secular trends and year-to-year changes in official national rates for homicide and motor vehicle theft are the least likely of all FBI Index crimes to have been substantially affected by increases or decreases in report rates during the post–World War II years.

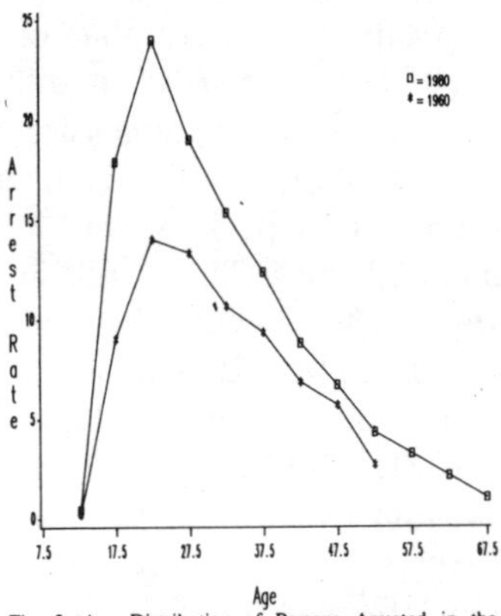

Fig. 2. Age Distribution of Persons Arrested in the United States for Nonnegligent Homicide, 1960 and 1980. For data sources, see notes 2 and 3.

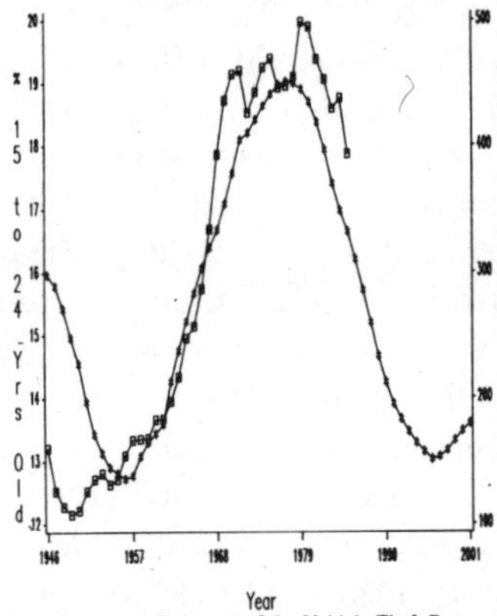

Fig. 3. Annual Estimates of the Vehicle Theft Rate and the Percentage of the Population Ages 15 to 24, United States, 1946–84, with Projections of the Latter to 2001. For data sources, see note 6.

FIGURE 7.16

The Two-Dimensionality of Graph

From Lawrence E. Cohen and Kenneth C. Land, "Age Structure and Crime: Symmetry Versus Asymmetry and the Projection of Crime Rates Through the 1990s," *American Sociological Review* 52 (April 1987): 173. Used with permission.

around it. We may reconstruct our figural version of knowledge in terms of enhanced "communication," whatever that may mean in the moment. But that assumes a representational and informational concept of knowledge—that which can be displayed and figured. We pretend methodologically that graphic work buttresses prose, which we believe is still constitutionally dominant in journal science. But that is backward; figure overwhelms prose with the promise of representation, of worldness, if not of photography itself.

Thus we "read" the graphs above as aids to prose interpretation. But they are more than that. They are texts in their own right, supported by words. Both graphs are surrounded by words—titles, explanatory notes about data sources, textual references. They do not, perhaps cannot, stand alone. Yet their significance is figural; they promise and seem to deliver science. Not only are data numbered but they seem to take shape and thus have meaning in graphic work. Graphs constitute ontologically when they give pattern to mere information, even quantitative information. Ontology promises the secret of Being; journal science discloses it figurally. We are assured of the orderliness and pattern of data by figural obtrusions in the text organizing the data into mountains, valleys, rivers—patterns.

As literature, the patterns in this figural work suggest a lawful world. Things fit together in a certain, discernible way. Figural methodology guarantees the conception of a nonrandom world, thus a nonrandom world itself. It does this in aid of communication, "visual aids" dotting the text merely serving to enhance interpretive understanding. Yet it is not clear whether graph or prose is the real text. A glance at the page suggests either. The trained science reader knows that figure is predominant and that text merely supports it, especially when it conceals its dominance underneath the pretense of literary convenience. Methodology reconstructs its figural work merely as ancillary—supportive, clarifying, efficient. Yet one can read and write data without forcing them into the mold of figure. Nothing about science requires that it be presented as the journals do. These gestures are only conventions—what make science seem scientific. But graphs are historical variations of textuality, not constitutionally essential. Mathematics organizes intelligence in a principled way. Yet, as Einstein showed, mathematics itself is a historical account, grounded in time, space, perspective—one might add, politics. Symbols remain just that—shorthand. But the more we believe that, the more they become the real text.

It is unsurprising that the data would be graphed in the illustrations above. The data involve age and vehicle theft rates; these are typically expressed in number and then arranged either tabularly or graphically. But in figure 7.17 graphed data address a topic typically not studied quantitatively, let alone figured. Here, the graph summarizes the rate of publication of utopian literature by year. Now if we assume, as I do, that figural ciples, norms, and rules of national and international conduct) develop during periods of hegemony and tend to collapse when hegemony breaks down. McMichael (1984, p. 5) pointed out that Great Britain was able to unify the world market during her hegemonic period with the ideology of political and economic liberalism. These basic principles become what Wallerstein (1980, p. 66) has referred to as the "untouchable ideological flagstones" that mark the boundaries of discourse during hegemonic periods.

The decline of hegemony is thus much more than a country's decline in economic and military superiority. It involves the breakdown of a qualitatively specific world order, including economic arrangements, relations between states and economic actors, and ideological justifications. This breakdown of existing dominant social relations and ideologies, coupled with the existence of new institutional arrangements in rising core states, should increase discussion of possible alternative social arrangements. Thus, our second prediction is that the number of literary utopias published in Great Britain and the United States increased during periods of hegemonic decline in the world-system. Conversely, we suggest that periods of hegemony, characterized by relative stability, should depress the number. Periods of relatively equal competition in the core should produce only a moderate volume of utopian literature as the scope of discourse narrows to discussions of the relative merits of the institutional arrangements of the existing competitors for hegemony.

Figure 1 shows trends in the rate of publication of utopian literature in Great Britain and the United States from 1883 to 1975.[2] Both countries display fairly similar trends during this period.[3] In fact, the zero-order correlation is .78. The similar pattern for both countries may reflect the effects of factors operating in a larger

[2] The publication rate for utopian novels is defined as:

$$\frac{\text{Number of positive utopias published}_t}{\text{Number of titles published}_t} \times 1000$$

where t ranges from 1883 to 1975. We did not include dystopias in the analysis.

[3] Figure 1 shows very similar general trends for Great Britain and the United States with two notable exceptions. During the periods 1888–1902 and 1930–36, the production of literary utopias is much higher in the United States than in Great Britain. These differences can be explained by the specific characteristics of each country. Roemer (1976) has argued that the rise of utopian literature in the United States at the end of the nineteenth century was a response to the closing of the frontier. This could explain the difference between Great Britain and the United States during the period 1888–1902. The higher rate in the United States during the period 1930–36 may reflect the fact that the economic contraction during this period was more severe in the United States than in Great Britain (Lerner 1975; Mitchell 1980; Rostow 1978).

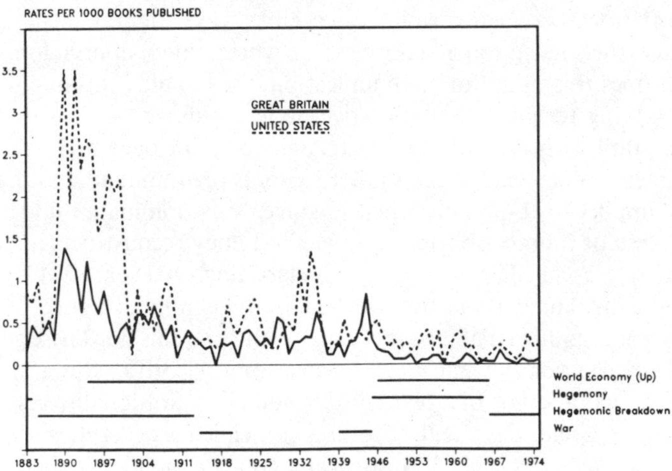

Fig. 1. Rates of Publication of Utopian Novels in Great Britain and the United States, 1883–1975

FIGURE 7.17
Graphic Work in Historical-Theoretical Analysis

From Edgar Kiser and Kriss A. Drass, "Changes in the Core of the World-System and the Production of Utopian Literature in Great Britain and the United States, 1883–1975," *American Sociological Review* 52 (April 1987): 288. Used with permission.

methodology is only a rhetorical device, then there is nothing inherent in the topic disqualifying it from this sort of gestural treatment. Yet it is clear that the authors use figural methods in order to legitimate a topic usually treated in different ways. They borrow figure from mainstream science in order to upgrade their own inquiry. This is not to say that they should not do so but only to observe the interest imbedded in their graphic displays. And it is not at all clear that they need the graph to do analytical work. But they use it anyway. In following the norms of journal science slavishly, they add value to them (even though their own political point of view might be different, as I suspect it is given the unusual nature of their topic choice and its authorizing theoretical framework).

When, then, can one legitimately graph one's analysis? Even the principle of textual necessity is ambiguous; one person's need might be another's frivolity. The question When? is answerable only in self-referentially rhetorical terms. One must recognize and then acknowledge that one's figural work is literary in nature, expressing a world to which one wants others to assent. One may not simply proliferate figures as if they epitomize science; one must argue for the uses of expression in turn arguing for one's worldview—and thus one's world. This is especially true today in analyses relying only weakly on mainstream methodology. Dissenters, in using the dominant apparatus, must foreground their reasons for doing so—particularly, to use the mainstream's tools against it, showing the different work they can do when liberated from a positivist representationality.

Unfortunately, few writers who use figure understand it as the rhetorical device it is. Instead, they photograph the world, thus recommending it, in exactly the same way as more self-conscious positivists. The graph in figure 7.17 tells little about the argument but a lot about the authors' view of mainstream science. The fact that it appears without explanation or apology deserves notice; clearly, outsiders want to be insiders. These things are difficult to disguise. Indeed, figural work might well occur in inverse relation to an article's mainstreamness (although what constitutes mainstreamness circularly involves the literary usage of that apparatus): The more marginal the article, the more it must rely on positivist conventions. This does not mean that mainstream articles do not table and graph their analyses but only that renegade writers must work that much harder to legitimate their work using the conventions epitomizing journal science. As I explore further in chapter 9, even "theory" writers figure their work, bending and shaping their prose into the convoluted patterns of journal science.

The world of journal science is not exactly monolithic—not exactly, but close. Writers are published who seem to use iconoclastic theoretical approaches. Marx is occasionally cited; feminist work appears. But they are the exceptions. And virtually everything published in mainstream science employs mainstream methodology as a literary device even if it seems to eschew mainstream theory. To have a hope of publication, dissident scholars must lard their texts with the rhetorical techniques of high science. As well, they must separate literature review from methodology from closing discussion. Intellectual pluralism is constructed by journal editors who claim to publish heterodox work. In fact, only the authorizing names are different. The methodologies—literary strategies—are the same. Writers clone themselves from cited authority and then figure their analyses, the twin devices of journal science. It matters little that the cited authorities occasionally include dead Europeans as well as live empiricists. Positivist representation is still intended by the work, especially in those cases—virtually all cases—where the authors use figural gesture to upgrade the epistemological status of their work.

There seems to be no escape; one must write what gets published. Even without an elaborate conspiracy, journal editors (academics themselves) reproduce journal science in its own image. Savvy writers "know" that one must conform methodologically to have a hope of publishing. And without published articles, tenure and promotion become pipe dreams. However much editors might wish to change things—and few do—things remain the same if only because we have been doing this kind of work for so long. Journal science exists as an exemplar of itself; to publish one must imitate the articles published last year and the year before that. In fact, this accelerates the trend. It is by now commonplace that one must do quantitative and figural work, and cite disciplined literatures, to publish in science journals. Old and middle-aged faculty inculcate this in younger faculty and graduate students. People at the cutting edge do cutting-edge work. Of course, this is tautological. Work that is "hot" reproduces yesterday's warm work. Thus if we saw numbered, tabled, and graphed work last year, we will surely see more of it. Writers are inherently other-directed; they write what they think will sell (and then, if they are professional, they convince themselves that this reflects them authentically).

Thus sociologists who study utopian literature quantify and figure their work unapologetically. They need not simply apologize, for survival is reason enough. Rather, they could interrogate the literary and political impact of their writing in a way that is consistent with their interest in utopian literature in the first place. Presumably, the writers in question favor utopian literature as a mode of social protest. But they do not protest their own disciplining. Instead, they reproduce it and thus everyone's. One can exonerate individuals struggling to survive in a world hostile to intellectuals without failing to see the implications of this for the larger intellectual and political scene. To trade in ideas today one must know methodology, hard ware, software, networks, literatures; these are the costs of survival. But, inevitably, the cost is too high. By writing in those genres without understanding them merely as literary and political conventions, one adds value to them, making it even less likely that next year's writers (let alone next generation's) will be able to resist.

In social science, the trendy academics are those able to parlay their theoretical marginality into mainstream methodology and thus flourish. They know how to play the game and win. They acquire real estate in the leading journals and thus jobs at the leading universities. Unfortunately, they also allow themselves to be claimed by disciplines trumpeting their heterogeneity. As well, they serve as moral object lessons for the young or unregenerate. "Look at so-and-so; s/he uses quantitative methodology, gets grants and has tenure. You can, too, if you do not let your foolish principles stand in the way." This is a conversation many (of the few) radical academics have had with chairpeople, colleagues, even their loved ones who must support them through thick and thin.[9] The price of defiance is judged too high, especially when the cost of conformity seems so low—familiarity with methodology as a literary logic of presentation. If only if were that easy! Instead, one is swallowed in one's work; intellectual unconsciousness is the outcome of intellectual betrayal.

I am not saying that tenure is not worth the effort; that is really beside the point here. Life under domination is less than a full life, thus occasioning numerous compromises. But it is an illusion that one can "use" positivist literary method without being used by it in turn. At the least, journal science requires a slavish obedience to authority that narrows one's topic and turns the prose into thoughtless picture. This is well and good if one can shed this skin after tenure. But simply publishing this kind of work makes it that much more difficult to publish other work, for it reinforces the literary folkways of journal science. In adding one's voice to the dominant approach, one is silenced and silences others in turn. Literary style is inherently political in this way; in encouraging others to replicate it, it forecloses defiant possibilities.

The graphic work in figures 7.18 and 7.19 moves beyond linear presentation; it is richer, deeper, denser. A number of figural rhetorics comingle in the graphs, connoting difference, overlap, intersection. Whereas the earlier graphs suggested two dimensions, these virtually suggest three.

It would take a specialist in Egyptian hieroglyphics or perhaps cryptology to crack these codes. They are not an easy read. Unlike the tables reprinted earlier in the chapter, these tables do not come with interpretive rules. One must make the best of the mess portrayed. This is an article relying on a somewhat unconventional methodology, content analysis, where data are newspaper articles coded into quantitative categories and then analyzed. Fearing impressionism, the authors work hard to add value to their scientificity; these graphs are the result. Their recondite code suggests a rigor and complexity to the study that the ordinary reader might not find in the prose. By being graphed, the data are raised in stature. Do the

824 AMERICAN SOCIOLOGICAL REVIEW

Figure 5. Black Unruliness, 1955–80

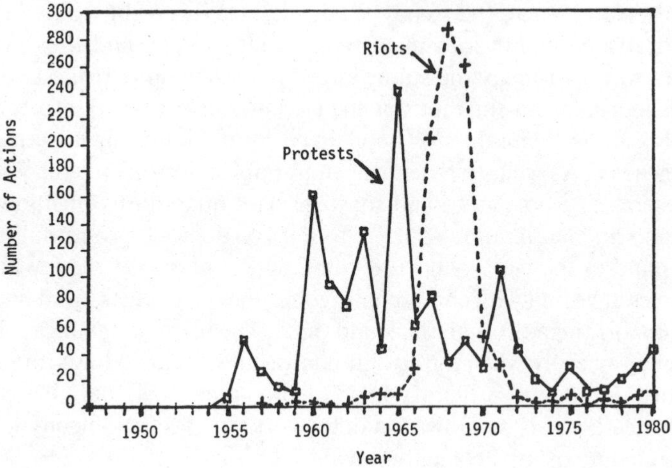

Figure 6. Major Issues in the Black Movement, 1955–80

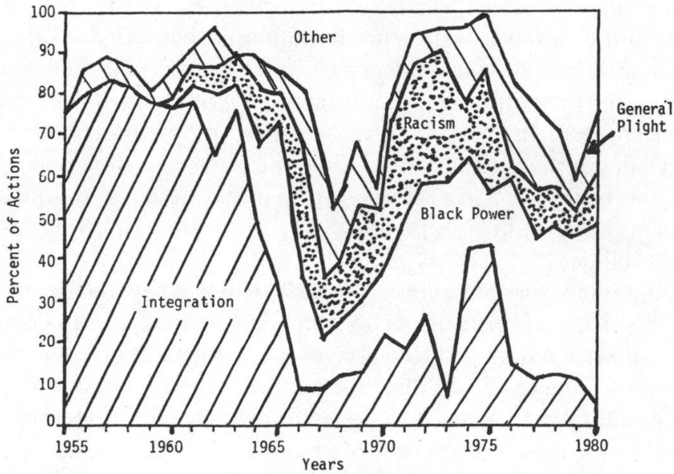

FIGURE 7.18
Dense Graph I

From J. Craig Jenkins and Craig M. Eckert, "Channeling Black Insurgency: Elite Patronage and Professional Social Movement Organizations in the Development of the Black Movement," *American Sociological Review* 51 (December 1986): 824. Used with permission.

Figure 7. Actions by Classical SMOs, 1961–80

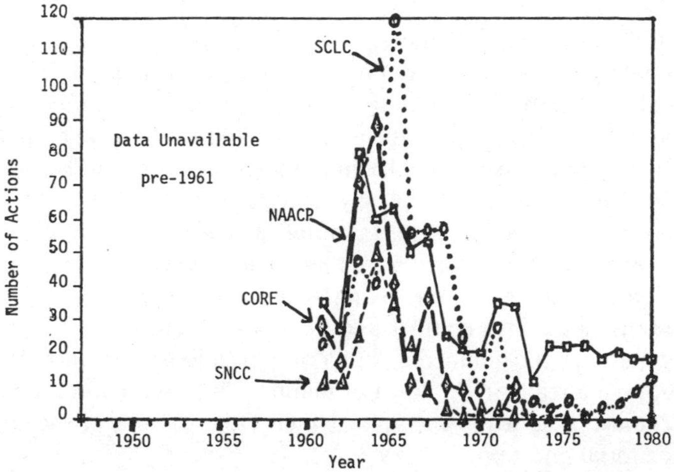

during the growth phase of the movement (1955–65). Likewise, crowds fell from 30.5 percent to only 12 percent of movement actions, rioting halted, and protests fell from over 200 per year to 10–20 (Figure 5).

Yet the social control argument also has problems in explaining the persistence of black power issues and unruliness. The professional SMOs apparently took their cue from the indigenous challenge. The shift to empowerment demands held through the 1970s. 34.9 percent of all movement actions focused on the older integration concerns while 30 percent addressed the new problems of powerlessness, 32.2 percent dealt with white racism, and 19.4 percent with the general black condition (Figure 7). Nor did the patronage blunt militancy. Although rioting disappeared, this was due largely to repression and ameliorative reforms. Protests dropped but they remained 23.6 percent of movement actions, virtually identical to the 25.8 percent of the tumultous 1960s. In other words, patronage may have accelerated the overall decline, but it did not restructure the movement's agenda or tactics.

Although the movement continued to score victories in school integration and employment discrimination cases, these were primarily a question of implementing prior victories, a set of tasks for which the professional SMOs were well suited. Black conditions did gradually improve during the 1970s (Farley 1976; Burstein 1985). Yet there were no great institutional innovations on the scale of the Civil Rights Acts of 1964 and 1965 or even the statutory strengthening of the Equal Employment Opportunity Act in 1972. In fact, the reversal of some of these gains in the early 1980s is eloquent testimony to the dangers of social movement professionalization and patronage. An excluded group cannot count on professional SMOs and elite patrons to protect and advance its interests against powerful opponents without sustained indigenous mobilization. If anything, this illustrates the central importance of indigenous organizing in creating elite patronage and initiating social reforms.

The greatest irony was the impact of movement decay on the patronage for the professional SMOs. Between the funding peak in 1973 and 1980, foundation support dropped from over \$3.8 million to \$1.4 million, a 62 percent decline. Although the shrinkage in foundation assets undoubtedly played a role, total foundation grants falling by 36 percent from 1970 to 1980, almost all the reduction came after 1975, suggesting that the major factor was the decline in indigenous actions. Ironically, the major losers were the professional SMOs, which dropped from \$2.87 million in 1973 to \$1.14 million in 1980. Over the same period, institutionalized organizations such as churches and universities, which had previously represented no more than 10 percent of movement

FIGURE 7.19
Dense Graph II

Source: See figure 7.18. 825. Used with Permission.

authors really need these figures to make a case? Again, the answer depends on where one sits. But it is hard to make the case that these graphs are as accessible as the ones above. Their oblique and busy representationality stands in the way of ready interpretation. As such, they embellish the scientific text, little more.

If the authors had submitted their article to a journal without this elaborate design, they might well have been rejected simply because their prose did not sport the usual signs of quantitative method, including elaborate figural work. They anticipated that and added these graphs. One might read figural work as a publishing strategy, elevating one's work in the eyes of editors where editors typically reproduce the conventions of figural expression in what they publish. I am not saying that nonquantitative articles are more likely to sport figure than quantitative ones, for number is itself a type of figure. I am saying that nonquantitative articles must translate their data into number and then other figural gestures precisely because they lack the numbered patina of typical journal science. If figure does not exist already in the form of number, then figure (and sometimes number) must be added by the savvy writer. Otherwise, he or she unwisely courts editorial rejection.

Manuscript reviewers and editors rarely comment directly on the lack of figural work, except perhaps to suggest that overtly theoretical articles might be propositionalized and then tested empirically. Figural work, including quantitative analysis, is presupposed as normative. Articles lacking such gestural embellishments are considered theory articles, even though for those of us who write theory, theory connotes something quite different—a digging at foundations or first principles, perhaps, or, as here, critique.[10] Journal science values figural work as evidence of rigor, even if reviewers and editors appraise the degree of rigor in different terms—say, lack of methodological sophistication. Methodological cavils by reviewers and editors spring from overall impressions of the work in question; one receives all sorts of signals about an article's scientificity—lack of number, table, equation, graph, model, picture. Workaday scientists and editors rarely recommend the addition of these literary techniques; if anything, they may recommend cutting figure in order to save space (at least in articles having nothing to prove about their scientificity).[11] Yet they notice them if they are absent. Methodologyless science is disqualified as nonscience; however one demonstrates the existence of methodology, one must. Thus unnumbered articles frequently resort to other displays of figure in purchasing the impression of rigor cheaply—assuming we agree about what constitutes rigor in the first place.

If some figure is good, frequently more is better; at least, habits die hard. We must carefully explain the abundant use of gesture in journal science somehow. The argument that gesture simply enables communication fails to address the principle of marginal utility. Too much design (along with too much complexity) presents interpretive problems of its own. Science articles do not intend to be intelligence tests. But the more they are cluttered with figure, the more they challenge intelligence to find examples of itself in the writing.

In figures 7.20 and 7.21 we have six graphic designs on two pages. On the first page, we have four graphs and only 25 half-lines of prose. That is about 44 words per graph—a meaningless figure except to mathematizing science. In this case, the graphs rest atop a variety of numbered formulations clarifying their data sources. Mud makes mud even muddier. On the next page, we find two more graphs but with more prose.

In the last case, we find a science page with five graphs, in this case relatively simple ones. Again, a rough check indicates about 44 or 45 words of prose per table. But the larger context of the entire article reveals page after page of semiotic presentation. The methodology section includes 10 pages of text and 18 separate figures (including one nearly full-page table). That is a figure rate of 1.8 per page. And the article (minus bibliography) is only 15 pages long. Thus every single page averages more than 1 figure; every single page averages more than 1 intelligence test, 1 hermeneutic puzzle, 1 code to crack. Perhaps this is a good way to present science. In any event, these figural gestures are not defended as the literary work they are. They merely obstruct a smooth reading and thus enhance the science aura.

Model

Some articles contain figural depictions of causal relationships, enhancing the causalities suggested in the so-called formal models specified in the methodology section. These depictions suggest one-way and multiple patterns of causality among variables studied. Whereas these data are frequently presented in the regression tables found in the model estimation sections of journal science, they are sometimes represented figurally, as in figure 7.22.

The figure summarizes the model specified or predicted in the first part of the methodology section. Arrows flow everywhere between and among discrete variables. If ever a case can be made for figural work, it is here; the model clarifies the sometimes obscure relationships postulated or described in formal or numbered terms. And if there is any doubt about the implication of causality in journal science today, it will be dispelled by inspection of these models. Causality is suggested explicitly, not just implied, in this art.

A closer reading of this causal model reveals the multivariate nature of much social science research. The postulated causality is complex; expected

Figure 1

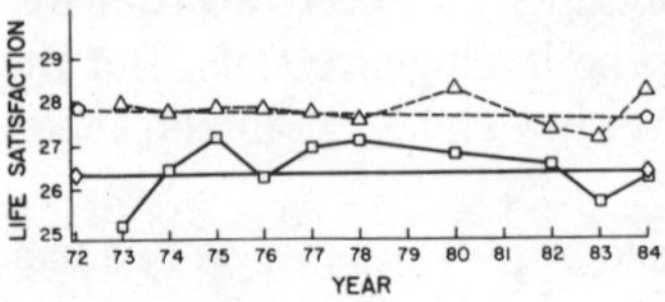

(1) Adjusted Means from the MCA Analysis:

Year	1973	1974	1975	1976	1977	1978	1980	1982	1983	1984
White	27.97	27.85	27.85	27.85	27.76	27.58	28.27	27.35	27.14	28.23
Black	25.15	26.46	27.18	26.24	26.95	27.10	26.78	26.52	25.65	26.22

(2) Selected Statistics from the Regression Analysis:

From the Regression:

b_0 (constant—$\hat{x}$ for whites in 1972)	27.85	
b_1 (year slope for whites)	$-.0237^{b}$	$^{a}=p.<.05$
b_2 (race effect—difference between $\hat{x}$ for whites and $\hat{x}$ for blacks)	-1.51^{c}	$^{b}=p.<.01$ $^{c}=p.<.001$
b_3 (year by race interaction—difference between slope for whites and slope for blacks)	.0267	
Derived Values:		
year slope for blacks	.0030	
$\hat{x}$ for blacks in 1972	26.34	

$R^2=.077^{c}$

Figure 2

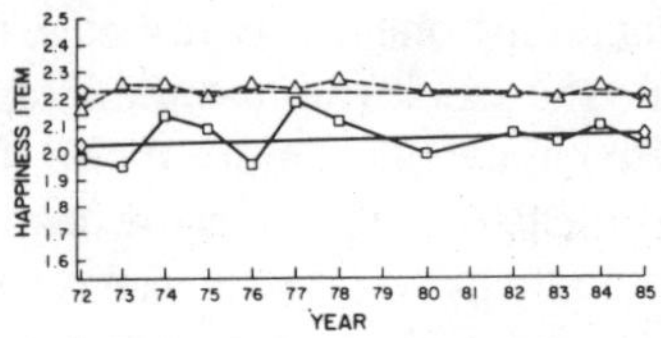

(1) Adjusted Means from the MCA Analysis:

Year	1972	1973	1974	1975	1976	1977	1978	1980	1982	1983	1984	1985
White	2.16	2.25	2.25	2.21	2.25	2.24	2.27	2.23	2.23	2.21	2.25	2.19
Black	1.98	1.95	2.14	2.09	1.96	2.19	2.12	2.00	2.08	2.05	2.11	2.04

(2) Selected Statistics from the Regression Analysis:

From the Regression:

b_0 (constant—$\hat{x}$ for whites in 1972)	2.2250^{c}	
b_1 (year slope for whites)	$-.0002$	$^{a}=p.<.05$
b_2 (race effect—difference between $\hat{x}$ for whites and $\hat{x}$ for blacks)	$-.1995^{c}$	$^{b}=p.<.01$ $^{c}=p.<.001$
b_3 (year by race interaction—difference between slope for whites and slope for blacks)	.0046	
Derived Values:		
year slope for blacks	.0044	
$\hat{x}$ for blacks in 1972	2.0255	

$R^2=.074^{c}$

Figure 3

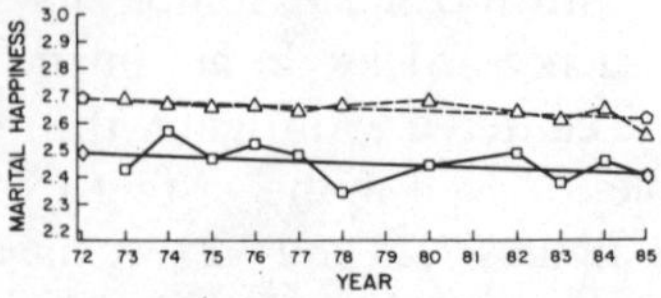

(1) Adjusted Means from the MCA Analysis:

Year	1973	1974	1975	1976	1977	1978	1980	1982	1983	1984	1985
White	2.69	2.67	2.66	2.66	2.64	2.66	2.67	2.63	2.60	2.64	2.54
Black	2.43	2.57	2.47	2.52	2.48	2.34	2.44	2.48	2.37	2.45	2.39

(2) Selected Statistics from the Regression Analysis:

From the Regression:

b_0 (constant—$\hat{x}$ for whites in 1972)	2.6856^{c}	
b_1 (year slope for whites)	$-.0075^{c}$	$^{a}=p.<.05$
b_2 (race effect—difference between $\hat{x}$ for whites and $\hat{x}$ for blacks)	$-.2003^{c}$	$^{b}=p.<.01$ $^{c}=p.<.001$
b_3 (year by race interaction—difference between slope for whites and slope for blacks)	.0008	
Derived Values:		
year slope for blacks	$-.0067$	
$\hat{x}$ for blacks in 1972	2.4853	

$R^2=.020^{c}$

Figure 4

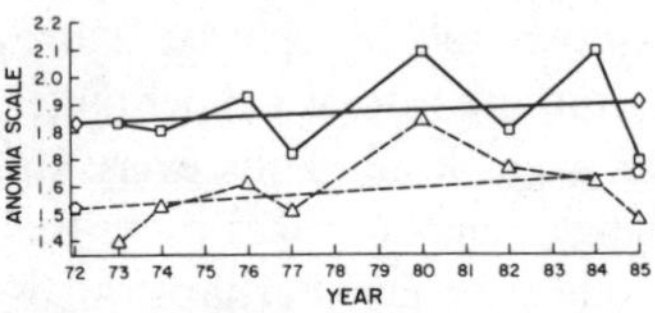

(1) Adjusted Means from the MCA Analysis:

Year	1973	1974	1976	1977	1980	1982	1984	1985
White	1.39	1.52	1.60	1.50	1.83	1.65	1.60	1.46
Black	1.83	1.80	1.92	1.71	2.08	1.79	2.08	1.67

(2) Selected Statistics from the Regression Analysis:

From the Regression:

b_0 (constant—$\hat{x}$ for whites in 1972)	1.5172^{c}	
b_1 (year slope for whites)	$.0093^{c}$	$^{a}=p.<.05$
b_2 (race effect—difference between $\hat{x}$ for whites and $\hat{x}$ for blacks)	$-.3152^{c}$	$^{b}=p.<.01$ $^{c}=p.<.001$
b_3 (year by race interaction—difference between slope for whites and slope for blacks)	$-.0047$	
Derived Values:		
year slope for blacks	$-.0046$	
$\hat{x}$ for blacks in 1972	1.8324	

$R^2=.113^{c}$

ness, marital happiness, trust, and physical health. There has been some variability in the 1980s in life satisfaction of whites, and great variability across the entire study period for anomie.

Changes in adjusted means for blacks do not parallel those for whites partly because the sample sizes for blacks are roughly 12 percent those of whites. Undoubtedly in some cases the differences between blacks and whites would not be significant if evaluated in a particular year; however, in no year does the MCA analysis show that blacks are better off than whites.

Overall the MCA analysis supports the interpretation given above of the regression analysis. Over the years 1972 to 1985 whites experienced significantly greater quality of life and psychological well-being than blacks, and the disadvantage of blacks showed no sign of statistically significant improvement over time.

Finally, it is important to emphasize that while we were able to find no evidence of convergence between blacks and whites in quality of life and psychological well-being, we

FIGURE 7.20

The Relationship Between Graph and Prose I

From Melvin E. Thomas and Michael Hughes, "The Continuing Significance of Race: A Study of Race, Class, and Quality of Life in America, 1972–1985," *American Sociological Review* 51 (December 1986): 837. Used with permission.

838 AMERICAN SOCIOLOGICAL REVIEW

Figure 5

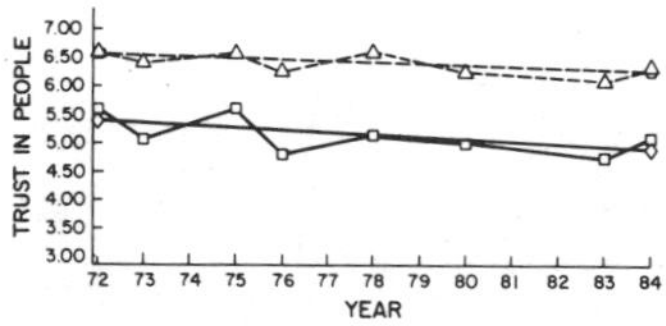

(1) Adjusted Means from the MCA Analysis:

Year	1972	1973	1975	1976	1978	1980	1983	1984
White	6.58	6.42	6.57	6.27	6.59	6.26	6.09	6.31
Black	5.59	5.07	5.62	5.81	5.13	5.00	4.73	5.08

(2) Selected Statistics from the Regression Analysis:
From the Regression:

b_0 (constant—$\hat{x}$ for whites in 1972)	6.54	
b_1 (year slope for whites)	$-.0249^c$	$^a = p. < .05$
b_2 (race effect—difference between $\hat{x}$ for whites and $\hat{x}$ for blacks)	-1.14^c	$^b = p. < .01$ $^c = p. < .001$
b_3 (year by race interaction—difference between slope for whites and slope for blacks)	$-.0168$	
Derived Values:		
year slope for blacks	$-.0417$	
$\hat{x}$ for blacks in 1972	5.40	
$R^2 = .141^c$		

Figure 6

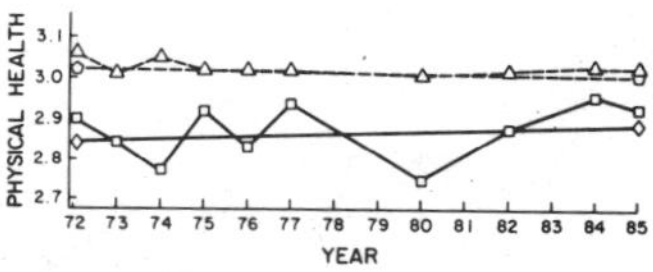

(1) Adjusted Means from the MCA Analysis:

Year	1972	1973	1974	1975	1976	1977	1980	1982	1984	1985
White	3.06	3.01	3.05	3.02	3.02	3.02	3.01	3.02	3.03	3.03
Black	2.90	2.84	2.77	2.92	2.83	2.94	2.75	2.88	2.96	2.93

(2) Selected Statistics from the Regression Analysis:
From the Regression:

b_0 (constant—$\hat{x}$ for whites in 1972)	3.0212^c	
b_1 (year slope for whites)	$-.0009$	$^a = p. < .05$
b_2 (race effect—difference between $\hat{x}$ for whites and $\hat{x}$ for blacks)	$-.1795^c$	$^b = p. < .01$ $^c = p. < .001$
b_3 (year by race interaction—difference between slope for whites and slope for blacks)	.0053	
Derived Values:		
year slope for blacks	.0044	
$\hat{x}$ for blacks in 1972	2.8417	
$R^2 = .173^c$		

also found no evidence of increasing divergence. Basically there has been no change in the difference between blacks and whites from the early 1970s to the mid-1980s.

C. *The Interactions of Age and Class with Race and Year*. It is Wilson's (1980) view that economic and social conditions are improving for middle-class blacks and deteriorating for the lower class. We can also conclude from Wilson's argument that race is increasingly less important as a determinant of life chances for younger blacks as opposed to older ones, as the latter suffered more directly from past discrimination. If Wilson is correct then perhaps among blacks the psychological well-being of the young and the middle class has been improving relative to whites and that of others has been getting worse. If such differences exist, thay are masked if we looked only at blacks as a whole. To evaluate these possibilities, we ran the regression analyses described above within categories of age and class.

1. *Age*. We divided the respondents into five age cohort groups based on the decade in which they turned eighteen years old. The relevant years are (1) 1970 and after, (2) 1960 through 1969, (3) 1950 thru 1959, (4) 1940 through 1949, and (5) prior to 1940. In the regression analyses run within these five age categories there are thirty possibilities for finding significant race × year interactions (five age categories times six dependent variables). Only the significant interactions are discussed below.

Over the 1972–1985 period blacks in the oldest age category (those who turned eighteen years old prior to 1940) became happier (p.<.05) relative to whites, while at the same time they became less trusting (p.<.05). Those who turned eighteen between 1940 and 1949 became more satisfied (p.<.05) relative to whites. Those in the middle age category, who turned eighteen between 1950 and 1959 became happier (p.<.01). Finally, those in the youngest age category, who turned eighteen after 1972 became less anomic (p.<.05). In no other instance was the interaction between race and year significant in predicting psychological well-being and quality of life.

Thus, only two of the thirty analyses show either that younger blacks improved in their well-being compared to whites or that older blacks got worse. Three of the analyses indicate that compared to older and middle-aged whites, the well-being of older and middle-aged blacks improved over the study period. In summary, there is no pattern to these findings supporting the idea that the well being of younger blacks is generally improving relative to comparable whites.

2. *Class*. In an analysis that parallels that in the preceding section on age, we divided respondents into categories representing different levels of social class. Using the three indicators of socioeconomic status from the analyses above, income, education, and unemployment, and measures of class origin (i.e. mother's and father's education; estimated relative income of family at age 16) both separately and in various combinations, we found no general tendency for the psychological well being of blacks in the middle class to have improved over the study period relative to whites. With one exception, no matter how we defined middle and lower class, we found that the effect of year for blacks was the same as that for whites regardless of class, and that whites consistently had better psychological well-being than blacks regardless of social class.

FIGURE 7.21
The Relationship Between Graph and Prose II
Source: See figure 7.20. 838. Used with Permission.

SUBURBAN STATUS CHANGE 271

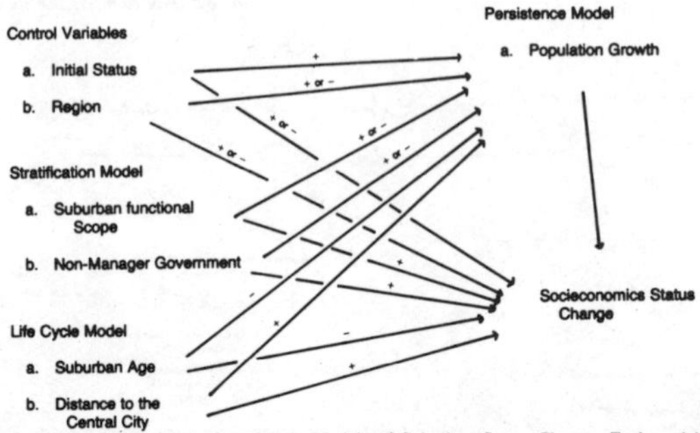

Fig. 1. Summary of Hypotheses for the Intradecade Models of Suburban Status Change. Each variable, with the exceptions of age, distance to the central city, and region, are measured at each time point and utilized in the appropriate decade model.

policies, tax rates, and provision of services.[3] Suburban governments can directly check status decline by, for example, zoning out lower-status groups (Shlay and Rossi 1981) by prohibiting the construction of low-cost housing. Zoning out the poor works well as long as a suburb is able to attract higher-status in-migrants necessary to maintain their status position. Population growth is not necessary to maintain or improve status in suburbs that restrict access by lower-status groups. Functional scope and accessible leadership are hypothesized to directly affect (positively) status change.

As suburbs become less attractive to higher-status migrants, they are forced to develop policies that encourage job and population growth. The basic assumption is that job and population growth will slow status decline by providing revenues to counteract the effects of aging and thus hold higher-status residents in place and perhaps attract new taxpayers to replace those moving out. On the other hand, a number of suburbs have chosen to adopt slow-growth policies (caps on new housing construction) to offset status decline. Growth from this perspective is undesirable in that a large population size is argued to influence negatively the quality of the environment through increased traffic congestion and pollution, poorer services at higher costs, and so forth. As a suburb becomes too large, it loses its ability to retain its higher-status population and attract suitable replacements. Suburban governments can influence status change by developing policies either restricting or promoting growth, depending on the individual suburb's situation and development strategy. As a result, it is difficult to make a specific prediction as to how functional scope and leadership accessibility will influence population growth—hence the + or − predictions in Figure 1.

In addition to population growth, age, distance to the central city, government type, and functional scope, initial status and region are included as controls in the models. Regional differences in status change and population growth have been demonstrated in several studies (Logan and Schneider 1981; Stahura 1979), and it is likely that they will occur in the present analysis. Denowitz (1986) found that initial status was related to status change. Hence, initial status in each decade model is hypothesized to be related to status change. Causal effects of initial status through population growth, consistent with the persistence model, may also occur; that is, initial status should be positively related to population growth (Guest 1978), which, in turn, should affect status change. The hypotheses in Figure 1 will be evaluated for each postwar decade.

[3] One study relevant to this argument is Schneider and Logan (1982), in which the impact of various policies on the growth of high- and low-income groups is measured. They found that service levels among communities have little influence on income-specific growth rates. They found, however, that high-income families tend to cluster in suburbs with high fiscal resources. This finding is consistent with the argument made here in that government seeks to reduce status decline by maintaining or improving fiscal status.

PROCEDURE

Suburbs included in the analyses are those with at least ten thousand population at the beginning of each decade. Sample sizes are 608 for the 1950s, 822 for the 1960s, and 1,163 for the 1970s. Use of the ten-thousand population criterion insures the availability of certain data

FIGURE 7.22
Figuring a Hypothesized Causal Model

From John M. Stahura, "Suburban Socioeconomic Status Change: A Comparison of Models, 1950–1980," *American Sociological Review* 52 (April 1987): 271. Used with permission.

correlations are low, explaining little about the causal contributions of each separate variable. As such, then, these models are pieces of artifice, unlimited except by the scientist's imagination. The arrows can flow in any number of directions. The choice of variables is inherently open. One can exclude interesting variables and possible causal patterns, just as one can include irrelevant variables. Of course, some of this comes out in the wash; model estimation confirms or disconfirms one's initial hypotheses. And yet social science correlations are notoriously low, explaining little of the "variance" in variables. There are no inherent limits to one's hypothesizing imagination, and ultimately there is little in the way of confirming evidence when correlations are so low. One can only redo one's causal model.

The model in figure 7.22 was subsequently tested against data. Other models exist wholly for heuristic reasons, suggesting possible future research. Frequently deriving from theoretical discussions, these models are included to enhance the science aura of the analysis in question. In figure 7.23, possible causal relations are sketched but not estimated with actual data, allowing one to assess causal strengths. This model suggests even more clearly than the previous one how poetic are the variables in question—how susceptible to reformulation, inclusion, exclusion. There is nothing wrong with this. Variables do not present themselves to the scientist in a pure state of nature; that is precisely my argument. Causal models implicitly suggest that the variables whose interrelations they sketch are real phenomena existing unadulterated in the objective world. As such, without being troubled about their overarching theoretical and metaphysical statuses, scientists can study these variables in their complex interrelatedness. It is as if the variables exist independent of their narrative construction by the scientist.

Many everyday scientists would puzzle over this. They would say that, certainly, variables exist; how else could we study them? More sophisticated methodologists might admit that variables do not exist as such but only point to underlying realities they seek to describe. Yet the figural work of causal modeling tends to conceal this literary understanding of the nature of multivariate analysis. We too often forget that the variables whose relations we sketch and then study empirically are fictions—thought products—whose reality we tautologically substantiate in the way we operationalize their observation and analysis. We can call anything "sun" and then define its observational nature and limits in certain ways. Eventually, we forget that "sun" refers to a real existent; instead, it takes on a life of its own. Our construction "sun" takes over for the real sun (whatever that might be); in studying it, we tend to give it false life, reifying it, as some would say.[12]

The analysis of variables can proceed as long as it remembers its fictive relation to what it examines. We conjure variables as pieces of reality out of the incredibly complex sentient world. In doing so, we risk forgetting—we

18 AMERICAN SOCIOLOGICAL REVIEW

vates interaction, whereas more sociological orientations underemphasize, but by no means ignore (e.g., Strauss 1959), these defense processes. As with the other controversies, there is little real incompatibility, since the hypothesis that pressures for the use of defense mechanisms will increase when self fails to be confirmed is probably acceptable to most interactionists.

Reviewing briefly the other motivating force in interactionist theorizing—the need to adjust, adapt, and cooperate with others—interactionism stresses the use of deliberative capacities for construction of definitions, orientations, and perspectives which can be used to map and construct lines of conduct (Blumer 1969). Such orientations circumscribe how individuals present themselves in situations, and depending upon the extent to which the reactions of others confirm these orientations, people feel a sense of adjustment and their fundamental needs for cooperation with others are consummated.

Figure 2 models these variables as they have been used to extend Mead's social behaviorism into interactionism. Two fundamental forces are hypothesized to mobilize human thought and action—the need for a sense of overall identity (Strauss 1959; McCall and Simmons [1966] 1978) and the need to cooperate, with the latter often influencing the former in creating needs for more situational identities (Stryker 1980, pp. 60–61). Such needs circumscribe mental deliberations of individuals by forcing the construction of (*a*) a conception of oneself revolving around configurations of substantive attitudes and (*b*) a set of definitions of, and orientations to, different types of situations. Self-conceptions circumscribe, to a degree, the construction of general orientations to situations. Moreover, one's self-conception sets into motion a derivative set of needs to maintain esteem of, and consistency among, the components of self-conception (Rosenberg 1979). Taken together, these motivating processes influence how individuals present themselves and interact with others. If confirmation of self does not occur, however, re-presentations of self are likely; and if self is persistently unreinforced, then the use of defense mechanisms becomes more likely (Horney 1950; Sullivan 1953). Similarly, if individuals do not confirm their definitions of situations, needs for cooperation set into motion re-thinking, re-definition, and re-presentations of self, but if individuals consistently fail to confirm their definitions of situations, then needs for cooperation (and indirectly, needs for identity) go unconsummated, thereby setting into motion defense mechanisms and/or the development of new self-conceptions, such as a deviant identity (e.g., Lemert 1951).

The Reemergence of Schutz's Phenomenology in Ethnomethodological Theory

While Schutz's work is often viewed as converging with Mead's, its origins reside in German phenomenology rather than American pragmatism. Yet, Schutz's (1932) early critique of Weber's action approach converted Edmund Husserl's phenomenological project into a more sociological theory of consciousness. Although Schutz's approach is similar to Mead's in its emphasis on the use of gestures and signs by individuals to put themselves in each others' place and although Schutz's concept of "stocks of knowledge at hand" reveals affinities to

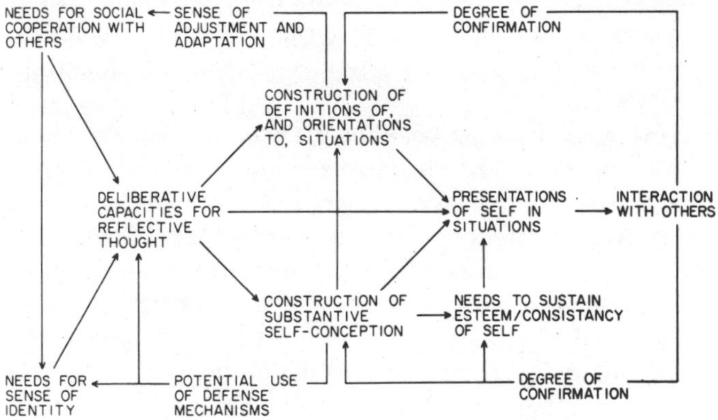

Fig. 2. Interactionist Model of Motivation

FIGURE 7.23
Figuring a Theoretical Model

From Jonathan H. Turner, "Toward a Sociological Theory of Motivation," *American Sociological Review* 52 (February 1987): 18. Used with permission.

often forget—that these variables and their implicit or explicit relations to other variables we construct are our creations. They do not unproblematically represent a given world out there but in fact help constitute it. The names we give to variables in their difference from other postulated variables reinforce the reality they aim to describe. We could as well come up with different names, different causal models, different worldviews; causal models are only images of a world reproduced by its namings.

Scientists learn this only to forget it. Too much metaphysical anxiety inhibits real research. But forgetfulness about the artifice-like nature of scientific discourse reifies, giving undue substance to what are merely narrative outcomes. We can write and act as if our variables exist; to do otherwise is nihilist. Yet we must not forget that variables figured in causal models are only as thick as the paper on which they repose. Like other aspects of methodology, they are merely literary techniques for conveying a certain conception of the world. In the model in figure 7.23 "needs for sense of identity" and "construction of definitions of, and orientations to, situations" recommend the world named in them. Perspectival and passionate at their core, these literary conventions allow analysis to take place. But they hinder it when we forget that we are their authors, not the other way around.

Picture

The final category of figural gesture used in journal science might be called simply *picture*. Occasionally the text includes photographs, but these are not the only pictures one finds. Not table, equation, graph, or model, pictures depict the actual reality to which they bear a supposedly representational relationship. They intend themselves not as semiotic translations but as reflections of things "as they are."

Figure 7.24 claims to describe "an interlock tie between two hypothetical firms." It promises a glimpse of the world itself, not merely its figural reconstitution in the indirect rhetorics of number and prose. Of course, no one has ever seen firms that look like this. There is enormous reduction and simplification going on here. Yet this use of figure is unusual in that it pretends not to mediate at all except in the trivial sense that it reduces the real to an appropriately publishable scale.

Figure 7.25 is another example of the use of picture as epistemologically constitutional. A portion of the world is depicted, mapped. Maps are simply figures claiming a one-to-one correspondence to the real; as such, they suppress their constitutionality, their contribution to making us think about (and thus live in) the world they recommend in their own image of it. When Americans who have not been west of the Mississippi think of Baja California,

782 AMERICAN SOCIOLOGICAL REVIEW

Figure 1. Stages in the Production of an Interlock Tie between Two Hypothetical Firms.[a]

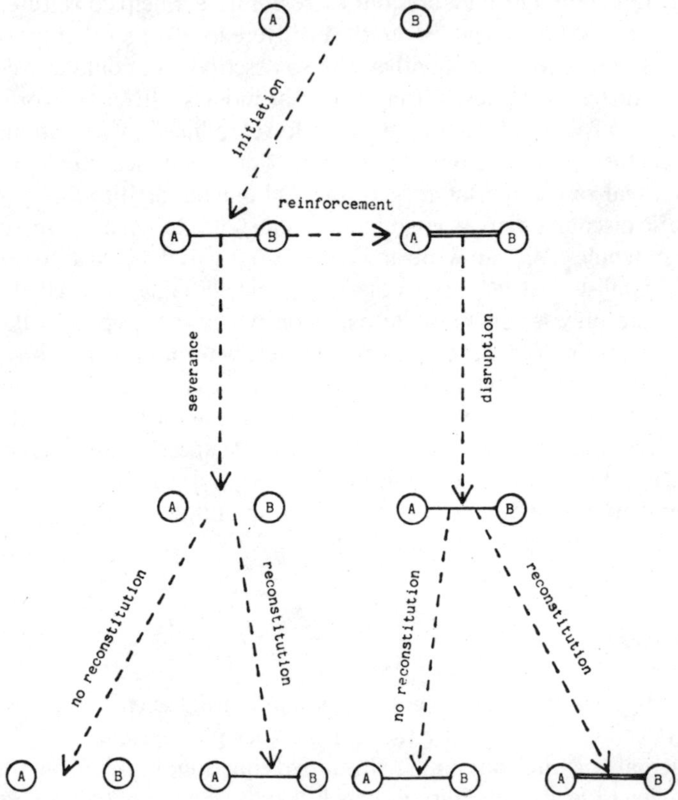

a. Circles represent corporations, solid lines represent interlocks and dotted lines represent stages in the tie production process.

The tie reproduction process has not been analyzed. While some assert that ties have a history (Mintz and Schwartz 1985), theories about the determinants of the interlock network's structure focus only on the initiation stage (Allen 1974; Pfeffer and Salancik 1978; Burt 1979; Pennings 1980; but see Galaskiewicz and Wasserman 1981). As a result, theorists may overlook determinants of network structure that operate at moments other than the initiation stage. Chief among them are the social relations in which ties may become embedded—the interorganizational linkages and intraclass bonds ties are hypothesized to facilitate. Further, while

FIGURE 7.24
Figure as Picture

From Donald Palmer, Roger Friedland, and Jitendra V. Singh, "The Ties That Bind: Organizational and Class Bases of Stability in a Corporate Interlock Network," *American Sociological Review* 51 (December 1986): 782. Used with permission.

72 AMERICAN SOCIOLOGICAL REVIEW

agricultural regions (Misner 1946; Henry 1937; Quicke 1966): the sandy loam, loam, limestone, and Ardennes regions.

The sandy loam region includes the southeastern half of Brabant province and the northern edge of Hainaut and Namur provinces. Such cities as Brussels, Louvain, and Tirlement belong to this region. The sandy loam region is one of the richest agricultural regions in Belgium and, while polyculture is widespread, market-gardening is the principal agricultural activity (Henry 1937, p. 9). Superior yields and proximity to urban markets make this region's agriculture highly market-oriented (Misner 1946, p. 159; Quicke 1966, pp. 79–84). Approximately two-thirds of the agricultural surface is under tenancy and one-third is under owner-cultivation. Nearly three-quarters of all farms are less than one hectare (Henry 1937, pp. 12, 16). Population agglomeration in this region averages between 490 and 1,700 inhabitants per km^2, making it among the most densely populated regions in Belgium (*Annexe Statistique* 1951).

The loam region of central Belgium covers Hainaut province and small sections of Namur and Liége province. The soil is deep and contains a productive mixture of clay, sand, and calcareous material creating Belgium's most productive agriculture (Misner 1946, pp. 60–61). The region specializes in the cultivation of numerous grains, market-gardening and forage crops, and industrial plants, especially sugarbeets (Henry 1937, p. 28; Quicke 1966, pp. 63–69; Misner 1946, p. 162). This region possesses a highly market-oriented agriculture because of its superior yields and its proximity to urban markets (Misner 1946, p. 161). Three-fourths of the agricultural surface is under tenancy with the remaining one-fourth under owner-cultivation (Henry 1937, pp. 12, 16). Like the sandy loam region, most farms here are less than one hectare. Tenants work most of these small farms, while industrial workers own and operate others (Misner 1946, p. 160). But large farms also exist in the loam region. Misner (1946, p. 161) notes that most of Belgium's large farms are located here. These farms frequently employ agricultural wage labor. Population agglomeration varies greatly in this region, with the southeastern zone averaging between 45 and 119 inhabitants per km^2 and the central zone averaging between 370 and 899 inhabitants per km^2 (*Annexe Statistique* 1951).

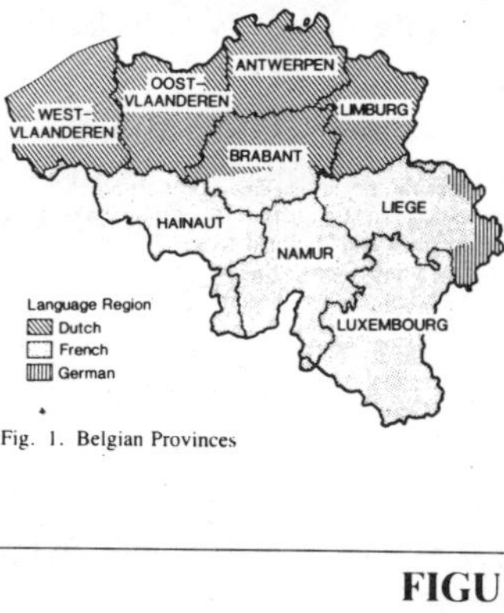

Fig. 1. Belgian Provinces

The limestone region is situated principally in the provinces of Namur and Liége, but includes a small area within the eastern part of Hainaut province. Its soil contains considerable limestone, but is nonetheless quite productive. The region specializes in cattle-raising, but it also produces some grains, alfalfa, red clover, and nursery products (Quicke 1966, pp. 45–47; Misner 1946, p. 162). This region's agriculture is also highly market-oriented for the same reasons found in the sandy-loam and loam regions, that is, high agricultural yields and proximity of urban markets (Misner 1946, p. 162). Again, like the sandy-loam and loam regions, small-tenant farming predominates in the limestone region (Henry 1937, pp. 16, 28; Misner 1946, p. 162). Population agglomeration averages from 70 to 179 inhabitants per km^2 (*Annexe Statistique* 1951).

The Ardennes region covers almost all of the province of Luxembourg and the southeastern third of Liége province. This agricultural region differs enormously from the other three. With an average elevation between 400 and 700 meters, this region is the highest in Belgium and the rainiest and coldest as well. The land is principally covered by forests. The soil is less fertile than elsewhere in Wallonia. The cultivation of crops is secondary; the principal economic activity is cattle-raising (Misner 1946, p. 163; Quicke 1966, pp. 35–45; Henry 1937, p. 28). More than in any of the other regions of Wallonia, the Ardennes' economic activity is geared toward subsistence, probably because of its low yields and distance from urban markets (Misner 1946, p. 163). While small farms and tenancy predominate elsewhere in Wallonia, medium farms and owner-cultivation prevail in the Ardennes (Henry 1937, pp. 12, 16, 27). Population is relatively dispersed, as measured by an average population agglomeration of 30 to 69 inhabitants per km^2 (*Annexe Statistique* 1951).

POLITICAL PARTIES AND ISSUES IN BELGIUM DURING THE INTERWAR PERIOD

During much of the interwar period, the Belgian political landscape was divided into right, center, and left. The Catholic Union stood on

FIGURE 7.25
Picture as Figure

From William Brustein, "The Political Geography of Belgian Fascism: The Case of Rexism," *American Sociological Review* 53 (February 1988): 72. Used with permission.

they may think of the finger-like spit of land they remember from grade school geography. Baja California exists for them in the reality of its picture. As such, then, the picture necessarily constitutes reality. In the absence of direct experience, the picture of Baja is Baja. Similarly, the relationship between hypothetical firms is those firms, at least as far as science is concerned.

We do not usually think of maps in this way. As pictures, they seem to make insignificant metaphysical claims. But cartography today is high science. We do not doubt that Baja the map looks somewhat like Baja the place. At least it would if seen from an orbiting spacecraft. But cartography yesterday was driven by superstition and untutored by direct experience. The world was flat, extending only to *terra cognito.* Beyond, monsters lurked, promising to swallow the intrepid if irrational voyager. But our picturing of the world is equally superstitious in the claims it makes on metaphysical trust. We assume that, for authorial purposes, the map or picture in question discloses enough about reality to be useful at the moment. The image of Baja tells us what happens when we drive south of San Diego, just as the image of interlocking firms promises what will happen when we become captains of industry. Whether or not these images are in fact useful is another question entirely.

The metaphysical claims staked by cartography seem to have few pernicious consequences. After all, the representation of Baja serves the purpose of getting there. Apart from the inevitable distortion involved in reducing the real to a picture, cartography enhances both instrumental and communicative rationality. By the same token, photography would seem to be an ally of enlightenment. "Telling" things as they "are," photography demystifies a world frequently idealized in prose and common sense. Yet photographic representation is an inadequate ontology, especially where science is concerned. In holding out the possibility of pure representation unsullied by perspective or medium, photographic realism promises a world that science simply cannot deliver. In pretending that it can and does, we mislead ourselves. Worse, we contribute to a world displayed as historically frozen—much less a problem when it comes to the depiction of Baja California.

Photographers know well that they constitute as well as reflect the world they capture in the quick literary work of the opening shutter.[13] To call photographs texts only strains the credulity of those who do not do or view photography aesthetically. Not only can photos be read, they read in their own right. By the same token, the map, if unaware of its own constitutionality, models an epistemology (and thus world) based on mapping as an adequate conception of knowledge. It is not, especially in the human sciences. Even picture postures, polemicizes, persuades; we must read it as metaphysical argument, especially when it suppresses or forgets its own constituting narrativeness as simply one version of the world among many. Artists know this; scientists frequently do not.

Journal science ignores these questions because the use of picture enhances its aura of authority and promises unproblematic representation. Picture summarizes science's ideal of itself: The text gains direct access to the world in displaying it. This is false not because picture may distort in the moment—badly drawn image—but because picture always distorts, neglects, selects. As such, picture is as fictive as any novel or poem. Knowing this, we can use picture playfully, passionately, and persuasively. Forgetting it, we delude ourselves that science (like cartography and photography) is cut from a different cloth than art. Saying this does not reduce science to art, but raises art to the level of science; it makes us rethink the metaphysical, hence political, claims we make when we act as if language discloses the world but that the world does not disclose (and edify) language. In figuring the world so, we help make it so, the secret ontology of positivism.

Chapter 8 Concluding Science

Normal Science Hedges Its Bets

Science must end sometime. Eventually it does. After the literature review, the methodology section, and report of the findings, articles peter out. They almost invariably end with "discussion" sections, frequently running less than a page in length. Pinched between the highly technical and figural discussion of findings and the proseless bibliography, the "discussion" evaluates the analysis in terms of aims delimited by the initial literature review and suggests further research. The discussion section—conclusions—has become a stock in trade of journal science. They all read the same. Their format seems scripted. They exude hesitancy about the results, careful to claim little for their analyses; they point up the technical-methodological limitations of the study; they call for further research, occasionally acknowledging policy implications.

The first notable thing about these "discussions" is that they rarely discuss in the ordinary sense of the term. They are not chatty, personal, impassioned, self-interrogating; they are dry as the rest of the text, albeit less larded with figured prose and figure itself. What they lack is more telling than what they include. Why are they not called "conclusions," as well they might be? Science is not to take narrative form, having a dramatic denouement; it is methodical, moving from its authorizing auspices through method and findings to a bare-bones discussion of the implications of the analysis in the light of what went before (and what may be yet to come in the way of future research). Journal science resists its reading as a narrative, a story, with a beginning introducing tension, a middle developing concept and character, and a concluding resolution of the initial dramatic tension. That is thought to be the stuff of literature. Science presents itself; it refuses to believe it also tells a story.

This does not mean we cannot read science as a story charged with authorial desire. I am trying to do that here. But part of the text is the text's concealment of its authorial motive. Calling the denouement discussion rather than conclusion further deauthorizes journal science. The less we can hear the interest of the author, the more authoritative the account will seem. The less intrusive extratextual motive in the text, whether politics or poetry, the more vouchsafed it will be as legitimate scientific analysis.

Discussions are called discussions for the same reason that journal science rarely writes in the first-person voice or interrogates its own desire autobiographically or parenthetically as part of the story itself. Science discusses, fiction concludes and resolves. The data, not the mercurial twists and turns of plot and fate, are supposed to determine the ending.

This explains why the discussion section is usually so short. While journal science runs from about eight to twenty printed pages—certainly not *Madame Bovary*—what is important is the ratio of discussion pages to the entire length and the relevant proportions of space occupied by the different sections of the paper. Invariably, discussions are the shortest part of the science paper. And they occupy only a small fraction of the total effort expended. Even literature reviews make bigger territorial claims than do discussions. This might be defended in a principle of authorial parsimony; good writing is spare and lean. But science is not concerned about being read as good or bad writing; its interest is representational. Thus discussion sections merely conclude data analyses; they do not enrich them. Indeed, discussions differ from the abstracts opening the articles only in the sense that they are printed in different type.

Apart from the hierarchy implied in length and voice—important issues in their own right—one must also consider tone. Discussions inevitably hedge their bets, ironically contradicting (some would say deconstructing) their postured representational nature. Whereas discussions act as if they merely summarize research findings, confirming or disconfirming stipulated hypotheses derived from opening literature reviews, they do less—and thus much more. It is striking how equivocal, even self-humbling, discussions tend to be. Tentativeness rules here, whereas attention to technical and figural detail dominated the methodology sections. Writers obviously hesitate to "go beyond their data," claiming too much for the particular analyses they have done. This is a fundamental literary problem for workaday positivism. Positivism aims to reflect the lawful universe, capturing its presence figurally on the science page, but it has metaphysical (thus literary) problems in extrapolating from the particular data at hand. This results in narrative ambivalence. In the discussion, authors evidence circumspection about what their analyses really signify in the way of these larger global patterns.

Science's literary ambivalence is usually kept under wraps. We assume that the authorial work deriving testable hypotheses from literature reviews, doing the research and then interpreting the findings proceeds straightforwardly. Little in the way of authorial temperament must be exercised to do this sort of work. Yet the discussions undo this suppression of temperament when it is immediately clear that writers are reluctant to draw precisely those larger ontological conclusions about the world authorizing their literary positivism in the first place. Few in the discussion section talk much about the contribution of their research to lawful understandings; the words "cause" and "causality" might be used sparingly in the course of the text, but no one talks about law, even though a representational concept of lawfulness animates everyday science. These are thought to be issues best left to philosophers, or at least to philosophers of science. As a result, the trivially narrow topics derived from literature reviews never broaden into more general knowledge about the world. This vitiates the very project of normal science, which is supposed to be cumulative.

Connections are not made nor are conclusions drawn in the discussion section. Instead, the authors bend over backward not to overreach their data in making more general inferences about the nature of the world. Science writers are ambivalent because they have been taught to view their work as contributing to the solution of a larger puzzle—the nature of the world. Yet they (think they) are methodologically prohibited from drawing these larger inferences. On the one hand, they seek laws; on the other, they prevent themselves from interpreting their data in a way suggesting these binding patterns. Of course, one might say that they are ill prepared to draw these larger conclusions given the rudimentary state of their method and theory. Scientists frequently say that about themselves, as we shall see. Yet this is fundamentally disingenuous, for people's research can invariably be understood to suggest larger insights about the world. Yet, for methodological reasons having to do with what scientists take to be technical limitations to inference, they foreswear these larger leaps—at least until the time that the methodological apparatus has improved sufficiently.

I think this is just an excuse. There is more involved than muddled methodology, as I say in the next section. Scientists are really at odds with themselves over their ontological mission; the issue is not methodology but metaphysics. Taught to be narrow specialists in their professional training, constrained by the literatures authorizing them, expert in using highly differentiated methodological techniques, science writers are unfamiliar with their philosophical roles. They do what Kuhn called normal science—piecemeal research—but they fail to contribute to its accumulation; the puzzle pieces never take shape because no one has the literary tenacity or gumption to do the shaping work. Instead, articles end with a whimper: It is said that little can be concluded from the research.

This puts representational science in a bind when it has claimed validity for itself and its multivariate methodology. Positivism promises knowledge of laws on the basis of which it can accurately make predictions. Yet it cannot fulfill its promise given what it takes to be methodological obstacles to sweeping inference. The most it can do is reflect, not reach conclusions. There is nothing about methodology per se that disqualifies these larger inferences; one can treat one's data in a suggestive and speculative sense, using them to essay hunches about the larger puzzle at hand. But science writers do not do this. Instead, they hedge when they attempt to suggest the larger picture. Trained to reflect the details, positivism cannot refract those details into a more syncopated whole. It misses the proverbial forest for the trees.

This is a trained incapacity. It takes narrative form when we examine the tentativeness of the discussion section.

> The historical evidence and quantitative findings presented above begin to suggest the promise of the authority-cost hypothesis as normatively-based approach.[1]

The data only "begin" to suggest; they do not suggest. Whether this is a slip of the tongue or just wordiness, it is not only that. On the one hand, according to positivists, the findings speak. They bank on their representationality. But on the other hand, the findings are ineffable. Data do not echo the general in their own particularity. The trivially narrow paper topics are further constrained by a literary strategy refusing to generalize beyond the data. Here, what is promised in the way of global knowledge is incredibly limited to begin with; the "authority-cost hypothesis" is not a major contribution by the looks of it. Yet the discussion hedges even here.

Certain of itself in the methodology and findings sections, science waffles when it comes to metaphysical work. The technical reason of the middle sections of journal science is not allowed to blossom into substantive reason insofar as the conclusion is concerned. Instead, science endlessly shuffles its feet, wary of overdrawing the local work it has done in the way of analysis.

> The analysis presented here is not a finished product. On the contrary, it represents one potentially useful strategy for reopening a serious consideration of a topic that has been submerged in sociological theory. And even if one does not accept the substance and strategy offered in this paper, alternative strategies should still move toward a sociological theory of motivation.[2]

Of course, the qualification imbedded in the first sentence of this passage is unnecessary. No analysis is a "finished product." But even less is claimed for the work than that. In the next sentence, the analysis is only a "potentially" useful strategy, not a useful strategy per se. Finally, even those who disagree with the analysis will agree with the salience of its topic, a sociological theory of motivation. Absolutely nothing is ventured in the way of definitive conclusions and absolutely nothing is gained. The writing is so ecumenical as to satisfy doubters, dissenters, and critics. A paper that begins narrowly ends even more narrowly, essaying nothing in the way of speculative interpretation or interpolation.

Indeed, what the discussions uniformly lack are interpretation and interrogation of one's results. Without them, the paper remains trivially descriptive and fails to speculate "beyond" the data (although with their aid). Again, science treats methodology as a real agent of epistemological constitution—of new knowledge—when in fact it is only demonstrative, a rhetorical device and nothing more. This is largely why the discussion sections of journal science contain so little in the way of imaginative extrapolation. Writers fear violating methodological canons of technique; they regard metaphysics as the plague, even if at some level knowledge is always metaphysical in its posits about the nature of the world. Journal scientists do not admit this; they view knowledge as indubitably representational, requiring them ironically to eschew the speculatively generalizing work promised in the canon of positivism. Representation by definition cannot essay itself into larger images of the world.

Thus science is confronted with two literary ironies, one at the beginning of the paper and one at the hand. In the beginning, science must authorize its analysis with respect to an overarching body of knowledge or literature. In establishing its fealty to canons, it hopes to derive the auspices for its own narrow analytical work. Instead, canon crushes it, reducing the paper to trivial iteration—at best, methodological innovation. At the end, science confronts another irony: In its slavish representationality, its grasp of the particular, it neglects the metaphysical *bricolage* (Levi-Strauss)[3] assembling the particulars into a general pattern. As a result, the cumulative process of scientific research never gets off the ground; each and every article ends in a ritualistic programmatism (discussed in the last section of this chapter).

Science confronts but does not resolve its second irony in the discussion section. Promising general understanding, positivist research cannot transcend the particular. To do so would be speculatively wild, unjustified by the data at hand. A language trick occurs that routinizes the metaphysical ambivalence of journal science. Instead of claiming positive findings (with presumably wider implications for theory and concept development), science contents itself to negate the negative: If nothing has been established in the way of positive knowledge, at least we know what is untrue. Of course negative knowledge is positive in its own right; through a process of exclusion, we eventually arrive at the whole truth. But these negations do not accumulate into their own negation, issuing in a new concept of the positive. Instead, they merely support interpretive apprehension as a methodological rule. Science couches its tendency to infer pattern from particular data in the methodological—really literary—injunction against overgeneralization. The only things we can know are the false; truth is being less wrong.

> The analyses presented in this article provide little support for the argument that educational stratification varies with the relative numbers of small and large families in the population. This issue, however, is not closed.[4]

Here, deconstruction reveals apprehension layered onto apprehension. The contention that existing knowledge is wrong is itself qualified by the possibility of its own error. "[T]his article provide[s] little support for the argument that." But "this issue, however, is not closed." It is crucial to read closely here. The claim that the "analyses...provide little support" does not vitiate them. It says only that we do not know if they are right or wrong. Science crafts discussions in this way, thus implicitly holding open the door for contrary evidence. The analyses may in fact be correct; that we do not know is held up as knowledge itself. But it is patent that knowledge claims *may* be correct. Everything and anything might prove out in the long run—even the claim that god exists. Here science, in avoiding metaphysics, invests metaphysically. Objective knowledge is really impossible. So, in effect, is refutation. We are left with the possibility of refuting refutation—a metaphysical possibility not requiring science to support it.

Science curbs its deeper desire to essay a concept of the universe. Thinking it avoids ontology, science opens the door to antifoundationalism. If refutation can be refuted, we lose the basis for adjudicating validity claims. Issues must admit of closure even if the arguments about them appear endless. Positivism denies the world-constitutive possibilities of reason because its own standard of reason—technical utility—is metaphysically suspect. Grounded in methodological desiderata, science's standard of truth depends on an external technology—progress in mathematics, computers, data sources. Opposing scholastic metaphysics, positivism in promising positive knowledge undercuts itself when it denies the possibility of closure altogether. When disconfirmation becomes the basis of epistemology, it opens itself to disconfirmation, quickly leading to the morass of relativism—the ultimate fatalism.

I want to read science's circumspection about drawing conclusions as constitutional, not merely contingent—a product of unreflected folkways (although it is surely that too). Representationality as ontology shortsightedly denies its own ontological premise; the reflection of the world disqualifies a conception of the totality within which history enters as the possibility of discontinuity as well as continuity. The technical—hence, philosophical and political—reason why science does not conclude laws is simply that laws do not exist in the social world. As agents as well as victims, people make and can undo the past and present. Freedom counteracts metaphysics while being metaphysically invariant in its own right. Reflection reflecting a certain lawfulness conceals its own contribution to what it represents as sheer fate. Thus it deconstructs itself. Science claims representationality in order to bring about a certain state of affairs. There is no such thing as mere representation.

If science admitted this about itself, it would suggest representation as a standard of knowledge only to disclose the limitations it imposes metaphysically. The representation of an unfree world reduces freedom even more. Thus science exists because there is too much freedom—too little obedience to patterns that science would reflect as laws. Science fails to draw large conclusions because to do so would involve metaphysical construction—the projection of the past and present into a determinate future. If science talked that way, it would lose political credibility. It is one thing to reflect imbedded patterns of inequality and individualism; it is quite another to put the name laws to them. At its limit, science in refusing metaphysics contradicts its own imaginative construction of the world as a horizon of determination. If science says "laws," it opens the question of its own nature as a text required to say it at all. For science metaphysically to constitute vitiates its own nonmetaphysical posture.

Everyday scientists understand their own circumspection entirely in terms of methodological injuctions against illegitimate inference. If one has not specified one's model to include a large number of variables and then collected data carefully on them, one cannot speculate legitimately about their possible salience. One can only talk about what one knows intimately to hand. Even then, science hedges; its strongest conclusion is that evidence does not exist to support a claim, not that the claim is wrong in the first place. This leaves open the back door for counterclaims in the future, saving reputations and ensuring duplicative research.

> Contrary to received wisdom in the social sciences and the mass media, we could find no evidence for religious secularization as measured by attendance at religious services in the United States over the past half century.[5]

This does not mean that religious secularization has not occurred. The evidence simply has not been found. Nor does it mean that religious secularization must be defined by attendance at religious services; other operationalizations are possible, reopening the question. This is as definitive as science gets in its discussion section. In general, the limits to inference are methodological in nature; it is hoped that one day better techniques and more reliable data will do the trick, allowing science to overcome its rhetorical circumspection in making bolder claims for its concept of the world.

Technical Limits to Reason

Science suppresses the irony that, as metaphysical constitution, it must draw back from definitiveness precisely in order to legitimate its representational work in the meantime. Or to put this backward, representationality seems to preclude a larger view of things, including the future. Either way, concluding as a literary practice both accounts for its own limitations in begging off more definitive interpretations—approaching lawfulness—and suggests future research as its complement. In this and the next section, I explore these literary practices. Whereas I understand science's aporetic disavowal of metaphysical constitution to be problematic in preventing science from essaying a larger conception of the universe, I also understand that science does not understand itself this way. Instead, science chalks up its metaphysical circumspection largely to methodological problems— issues to be resolved in future research.

Of course, this is predictable. Science would understand its own limitations in terms of what it lacks in the way of method, data, computers, replication, time series, and so on. This way, there is less to fix; better, fixing it—inadequate methodology—diverts it from larger metaphysical and hence political questions. It is commonplace to science readers that technical problems constrain the analyses at hand; interpretation is confounded by technical limits to reason, not by the avoidance of reason itself.

> Definitive statements about cause are hard to draw on the basis of a single nonexperimental study (Cook and Campbell 1979).[6]

I would say that "definitive statements about cause are hard to draw"—period. This is so for metaphysical reasons: People are free to disobey the causes. One day they may even overthrow them, reducing causality to prehistory. But science defends its interpretive hesitation on methodological grounds: Technique has not advanced far enough, budgets are insufficient, computers are inadequate, too few similar studies have been done, equations are insufficiently specified. Whatever the particular reason for science's ontological hesitation, the reason is particular.

> We urge caution in accepting our results and interpretations. While our research design is superior to previous studies in most respects, and the number of domains of behavior examined is extensive, there are important limitations...[7]

These limitations are technical in nature; they involve research design, not reason. In demystifying science's reasons for metaphysical equivocation, I am not denying that methodological problems may hinder insight. Indeed, as all practitioners recognize, it is possible to lie with statistics—distorting, deceiving, dissembling. Only sometimes is this intentional; other times it simply creeps into analyses resting on inadequate methodological translations of concept and hypothesis into researchable terms. But methodological issues do not decide epistemological debates, for they are only rhetorical devices used to persuade and recommend. These debates are decided by reason—circularly begging the question What does reason mean? Answers to that question themselves have to be carefully and powerfully reasoned; there is no escape from the narrative requirements of rhetoric.

Representation will never stand voicelessly on its own. Technique does not hum along noiselessly; instead, it requires human articulation, explanation, supervision. High technology duplicating itself in high methodology does not replace speculative philosophy—reason—as the arbiter of intellectual disputes. The logic of control is not disinterested; indeed, the pretense that it is, is deeply metaphysical—perhaps more metaphysical even than metaphysics in that it allows no room for dissent. At least, traditional speculative reason recognizes its own inherent finity; in needing to speak, it realizes it could speak differently. Democratic ontology completes itself in dialogue. Thus methodology as such solves nothing; it only postpones the search for more fundamental interpretations, speculations, imaginings.

Sometimes, methodological limits to reason are mundane.

> Our lack of data on submissions and evaluations of manuscripts does not allow us to test these alternative hypotheses, but the topic is worthy of future research.[8]

The reasonable if methodologically unschooled response might be—well, collect more data so as to probe these "alternative hypotheses." It is not enough to recommend research for others. I am pretending that closure is possible; it is not. But when one avoids probing alternative hypotheses simply because data are missing, one guarantees weak conclusions, or, as is more often the case, no conclusions at all. If data are impossible to obtain—too costly, too time-consuming—then abandon the project at hand. Technical innovation is insufficient reason to do research.

For nonpractitioners of journal science, it is sometimes incredible how self-effacing science will be. Writers do cartwheels to show their methodological—hence, intellectual—humility. It seems as if nothing went right with the research in question.

> Furthermore, the data and results reported here are limited in many respects, particularly by being so distant from the workplace relations they attempt to capture.[9]

> To be sure, the effects of gender mix on well-being were not statistically overwhelming, and the small numbers of men in mixed and female-dominated settings warrant caution in drawing strong inferences from these results.[10]
>
> Finally, there are limitations that derive from the nature of the sample.[11]

If the work is so problematic, why publish it? Cynics would suggest that journal science trivializes itself in order to leave more work for itself and others to do. Interpretive hesitancy makes room for other interpretations. And careers do not necessarily suffer if scholars reverse their field or are reversed by others. Frequently, the most reputable academics are those most embroiled in controversy; their work gets noticed, argued, cited. It matters little that they have changed their views. Indeed, intellectual self-correction is regarded as evidence of flexibility, maturity, growth.

However banal or profound the authors' caveats, it is notable that virtually every article includes some evidence of circumspection. These acknowledgments of possible error sometimes constitute the bulk of the discussion section; other times, they are mere epigrams dotting the conclusion with the residue of self-doubt.

> In considering the import of our findings, one must keep in mind the limitations of the analysis. First, the data are not longitudinal... Second, we lack data... Third, our best identified models... Finally, there are limitations that derive from the nature of the sample.[12]

The phrasings are too similar to ignore. The self-reported "limitations of the analysis" becomes a catechism, muttered with bent head amid thoughts of redemption. The discussion section plays a liturgical role, demonstrating the sacred nature of the foregoing analysis. It is unseemly to revel in the profundity of one's analysis. Instead, one must trim one's sails in a show of humility. This becomes a convention, not a considered literary gesture. Science, to be science, must admit its methodological limitations, refusing to make unwarranted metaphysical claims and sponsoring future research—technical innovation. Of course, the acknowledgment of limits dialectically makes way for other articles. Other versions correct and thus—or so it seems—they are solicited. Humility is simply a license to proliferate accredited disciplinary work.

I seem to be saying that analyses are always more significant than the discussion section claims for them. Frequently, though, that is not true; the liturgical self-humiliation at the end of the article is justified on the evidence. The article promises little, and delivers. Most articles afford few insights and make way for little in the way of broad-gauged speculation. Instead, they make tiny technical headway, if that. They are because they exist. They figure because they contain figure.

> For these couples, the amount of time together seems related to marital quality, but the kind of time together also matters: the more time together in activities such as eating, and playing, and conversing, the more satisfying the marriage. Unfortunately it was not possible to unravel the direction of possible effects. Results indicated the presence of a relationship, but whether it is recursive or nonrecursive is impossible to determine.[13]

Here, an article fails to make good on its promise; causality is "impossible to determine." The strongest conclusion is that one variable "seems related to" another, nothing more. I am not faulting the research design or interpretive approach. But one wonders if such an analysis has a right to exist at all, especially when positivism promises lawfulness, nothing less. Of course, much of journal science is exactly like this: Causality is not established. At best, relationships (between variables) are reported. At that, the authors do not openly admit their failure; instead of talking about causality plain and simple, they use euphemisms: "direction of possible effects," "recursive or nonrecursive." A metaphysical shortcoming is turned into a mere technical limitation. Presumably, more and better methodology would have plugged the leak. I doubt it. But if one accepts the possibility of causal analysis of the social world, one at least must deal in those terms directly. Philosophical problems do not disappear because they are called something else. Even if one has not established causality because—as for me—causality does not exist, one is enabled to claim it as a speculative venture; at least, much journal science implies a knowledge of causality even if it does not say so directly. This article is simply more candid than some.

> Our findings therefore are not all of a piece. They do affirm one simple conclusion, however: Marxists and non-Marxists alike will benefit from operationalizing the concepts of class, gender, and delinquency as we have done in this paper. Indeed, our findings indicate that to do otherwise is to obscure the class structure of gender and common forms of delinquency.[14]

In this instance, the authors conclude with moral exhortation: One should "operationaliz[e] the concepts of class, gender, and delinquency." The authors admit that the "findings are not all of a piece." They mean that not much has been learned. The payoff is strictly methodological, allowing theoretical categories to be measured—operationalized, as the jargon has it. I do not disagree that one might apply concepts to the empirical world; obviously, I do the same here. Yet one does not need many journal pages to establish this. In fact, the authors' analysis has revealed a mixed picture, making clear-cut interpretation difficult. At best, they have engaged in a methodological exercise, demonstrating the utility of what they call operationalization. That is well and good. But they could have dispensed with the apparatus of positivist inference and simply treated this methodological issue as rhetorical in nature, suggesting a certain literary usage (call it operationalization) in future science.

But they do not, for they are subservient to their "findings," the semiotic artifacts of a particular research design and methodology. The findings do not voicelessly suggest that "Marxists and non-Marxists alike will benefit from" certain literary usages any more than it shows that they should reduce their consumption of cholesterol. The authors believe that a particular measurement of class, gender, and delinquency is rhetorically useful, yet they infer that contention from their "findings." But findings do not find anything; they are found and formed as speculative constructions. Scientific measures serve exactly the same function as certain literary devices from free verse to onomatopoeia: They serve to illustrate, enlighten, explain.

We do not usually regard literature as science or science as fiction. We miss the validity claims imbedded in fictive narrative, just as we miss the literary artifice of science's text. This is because the notion of the "two cultures"—the humanities and science—by now is second nature. We make truth the business of science whereas literature indulges in fanciful illusion; as such, the humanities are moral and spiritual boons in a godless age. Yet both discourses intend validity; they merely go about it in different ways. Science is merely a component of literature, differing in that it must claim validity explicitly. Both convey, even recommend, certain versions of the world. They are products of literary artifice, stylization, arrangement; as such, they could be different. The best writing, in recognizing its finity, solicits other versions as a good community itself, the babel of writers offering a vivid example of what Marx called social freedom.

Science promises the truth, authorizes itself by showing its filial devotion to an established canon, and guarantees representation by figuring the world it addresses. When it fails to deliver, it makes methodological excuses. It is assumed that finding nothing is actually to find something—the world does not work as we or others imagined it did. Sometimes that can be important. But when virtually every science article ends with methodological apologies, something is going wrong. Writers are writing simply to write, laying claim to the valuable real estate of mainstream journal pages. They guarantee that their topics are trivial by deriving them hagiographically from existing bodies of disciplined knowledge; they excuse the triviality of their work by offering methodological caveats and excuses.

> [W]e lack data about many pertinent aspects of the classroom, school, and school district, and so we cannot fully explicate the effects of school structure on either educational self-direction or personality.[15]

Instead of "fully explicat[ing] the effects of school structure on either educational self-direction or personality," the authors accomplish a trivial and inconclusive analysis in its stead. Again—as always—methodology is the issue; reason is limited by technical constraints. While it is fine to express one's humility in face of the immense unknown, one hopes for more than that. At best, science articles make minor technical innovations—at best. And even these are couched in hesitating tones.

> Despite these limitations, a few factors give us some confidence in our measure of gender composition and the pattern of results obtained.[16]

Limitations are acknowledged—technical ones. And the authors can cite only a "few factors" giving them "confidence" in their analysis. But the point of the analysis is not analysis; rather, they attain only a better "measure," small payoff indeed. I am certainly not against methodological innovations; I dispute that methodology is the epistemological problem and that more or better methodology will set it right. The problem is the lack of reason, not the immaturity of technical reason. In blaming methodology, science writers burden it with too many expectations (thus giving themselves and others like them future work to do). Humility conceals a deeper arrogance about the worth of these forays in the first place. Meanwhile, intellectual progress does not take place; regression describes not a statistical technique but a civilizational tendency. While journal science cannot be blamed directly for the world's miseries, it is culpable to the extent to which it does not do more fundamental work—and here I do not require science explicitly to tackle "social problems" but only to understand that real intellectual interrogation facilitates liberation at a time when the rate of intelligence continues to fall.

Whether or not journal science explicitly confronts war, poverty and discrimination, it should address substantive intellectual problems especially when academic work is incredibly—and increasingly—trivialized. Methodology as metaphysics stymies more fundamental metaphysical, hence political, forays. People, instead, could be inquiring into the fundamental nature of things—the good, truth, beauty, justice. This need not take place within disciplinary philosophy; today that is among the least likely places for it to happen. Yet good intellectual work is not trivial; it asks and tries to answer big questions about our mortality and how it is that we spend our short lives badly. If this is political work, so be it. Conservatives and radicals join forces here; both applaud Greek questioning about fundamental issues, if not Platonist solutions.

This need not disqualify empirical research; indeed, such questioning today would bear directly on real-world concerns. Nor do we have to abandon quantitative method and other figural gestures as the tools of the oppressor; they are tools of oppression today, but they could be liberated, poetized. If methodology is only metaphysics—discourse, art, fiction, politics—then it legitimately claims a place in the pantheon of knowledge. As a fetish of technique, however, methodology neither poses nor solves intellectual, hence political, problems. It generates a huge research agenda for itself, promising advance but never registering it. As science becomes technique, it loses its mind; a casual perusal of journal science today reveals this to be so. Not only do we waste forests in proliferating arcane journal science read by virtually nobody; we waste time better spent in more fundamental intellectual activities, whether these take empirical or theoretical—or literary—form. At root, every discourse is poetry and science at once. It claims objectivity yet arises from the passionate subjectivity of a writer who writes as recommendation. I would rescue methodology from its fetishism and at once rescue theoretical inquiry from marginality. Eventually, the two could meet.

Normal Science as Programmatism

> Males are often said to experience strong peer pressure to engage in coitus. Nothing in the present analysis contradicts this...We pose the following hypotheses for further research...[17]

This concluding passage starts in the usual way. A disclaimer is offered; given the rudimentary state of knowledge, the most that can be said is that "nothing in the present analysis contradicts" the conventional wisdom about what leads young men to want sex. Apart from being trivially obvious, the claim is supported only through indirection. The authors say that it has not been refuted. This is by now a familiar device with which science appears (but fails) to avoid metaphysics. Yet below it we find an interesting construction, a literary transition of sorts that occurs in numerous science conclusions. After hesitating confirmation or disconfirmation is tersely cited, science engages in programmatism, calling for "further research" to settle intellectual questions. If the truth be not known, more science is called for.

More than a routine gesture, the programmatic nature of science conclusions buttresses science's metaphysical hesitancy. By acknowledging that the results are not yet in, we beg credibility for our own limited ventures. In delimiting our claims, we allow others to shoulder the ontological burden. In soliciting future studies, we act in a professionally responsible way, deferring to the community of science, where it might correct our own limited analyses. And, not least, programmatism keeps us and others like us busy; the less we can venture metaphysically, the more gainful work we do in the meantime. Above all, science acknowledges the possibility of its own correction in order to get in on the action.

Programmatism is more than simple convention, the parting shot of disciplined professionals. It is that too. Science's credibility rests in large measure on its participation in a community; thus anchored, each piecemeal analysis adds to the one just before it. Eventually, it is hoped, the big picture emerges; science discloses a lawful universe. But that does not happen. Instead, trivially narrow articles fail to connect; there is no accumulation of insight, evidence, understanding. At best, people vouchsafe existing disciplinary traditions within which they plow their narrow strips of the field. Authority is thus reproduced authorially. As a text, programmatism displays appropriate professional humility. The science writer matters less than the evolving body of knowledge to which each analysis makes its minute contribution. As a subtext, it further reduces intellectual work to trivial technical and methodological concerns, apologizing for the inconclusiveness of the analysis by suggesting more fundamental work for others.

As a literary practice, programmatism exhibits some common features. First, science simply registers the requirement of additional intellectual work; this is matter-of-fact, unargued, merely noted.

> Research is needed...to identify the circumstances under which...[18]
>
> Our results suggest several interesting avenues for further work...Kanter, for example, argued that... Martin has elaborated the relevance of these ideas for...[19]
>
> We hope our ongoing longitudinal analyses, as well as future research by other scholars, will indicate whether there is reason to expect similar patterns of...[20]

It is assumed that analysis provokes reanalysis and new analysis. In fact, this is an exceptionally deconstructive understanding of science's narrative nature. Each science article is a version, suggesting, even provoking, other versions. This insight could be the basis for rejecting a representational epistemology, as I would recommend here. But, instead, science suggests a continuity to itself that is false on the evidence. Study does not beget study smoothly but lurches from one narrowly unconnected topic to another.

While there is continuity in the literature people cite to authorize their research, there is virtually no continuity in the empirical results. Articles peter out with methodological caveats and apologies, doing little to move the authorizing literature forward (or in a different direction). Science postures that each analysis unproblematically suggests additional research building on it. I would contend, differently, that each analysis is only a version of versions, requiring response and dialogue both as correction and community. My version of science does not call for further research but simply expects it, suggesting the question to which it is a corrigible answer.

In the second passage above, the data are said to "suggest several interesting avenues for further work." In fact, data are voiceless; as literary embellishments, they are surrogates for authorial voice. The suppressed writer speaks through his or her "results." As such, the author recommends subsequent lines of inquiry; the data do not require them. The logic of inquiry is not driven by gradual disciplinary accumulation. Instead, it is moved—forward, backward, around—by authorial will. The first-person writer prefers to let his or her "results" speak for him or her precisely because it is understood that science is not authored at all. As representation, it lacks the curiosity or imagination to "suggest...interesting avenues for further work."

In the third passage, the authors assume that "future research by other scholars" will be carried out. Indeed, they know that their line of inquiry will have a future because they are already doing "ongoing...analyses." Whatever defects the present analysis contains, they will be corrected—by the authors themselves; indeed, they *are* being corrected in "ongoing" work. One wonders why they rushed to publication. Was it simply impossible to include a sense of their "ongoing" work in the space where they write their conclusion? Probably not. Instead, they save their thunder. Their future work gains authority from the present one. While all of one's work is intertextual in this sense, persupposing what went before and suggesting what may yet emerge, the authors here are saying more than that. Not only does the present analysis suggest new lines of inquiry but future work will address the same topic. The next science paper by the authors will not be different but the same; it will "indicate whether there is reason to expect similar patterns of" something or other. In effect, they excuse the shortcomings of their present work by promising fundamentally similar work. How strange.

One is tempted not to read so deeply into these lines, probably penned quickly to satisfy convention. Without saying these things, one risks being read as a person with all the answers or, worse, with the last word. And they help make way for their future ("ongoing") work by referring to it. Yet we must read deep into it for it is there—saying something to some readers, disgorged from the depths of some long-since-repressed authorial desire. Freud is right as far as he goes; everything has meaning. As deconstructors argue, too, nothing has meaning as well—literally, no-thing: absence, silence, blanks. We must take every text seriously, and even some texts appearing not to be texts at all, such as figure. Not to do so is irresponsible, further degrading language into cliches: "Research is needed..." Meaning should be given to words when they have lost it. Language cannot be opposed unless it is made to speak.

Beyond these concluding cliches intending deeper significance, one sometimes finds passages requiring future work to take account of them. As such, science claims its own authority, attempting to install itself in the disciplinary stock of knowledge from which it originally derived its own authority to write.

> Hence, future research attempting to explain patterns of intergroup criminal victimization would do well to attend to the structural opportunities for interaction between members of different social groups.[21]

Science heralds its own significance by anticipating how it will become part of its discipline's literature. This analysis claims to make a contribution ignored by future scholars only at their peril. Of course, if read carefully, the passage above only implies that it has met the requirement it imposes on "future research." Has it "attend[ed] to the structural opportunities"? In any event, it makes its point; its moral authority derives from its advisory, even cautionary, tone. Who knows what will happen to people who fail to "attend to the structural opportunities," heedless of the admonition "would do well to"?

On even closer inspection by people trained to read through argot to its essentials, the writers recommend nothing particularly surprising. In fact, what they suggest is little more than a banality. The "structural opportunities for interaction between members of different social groups" is made to sound more technical—hence, more profound—than it is. "Structural" means built-in, given, or fixed. And "members of different social groups" does little analytical work here. Thus the authors claim intellectual authority by threatening future science with illegitimacy if it ignores what is patently obvious. I do not deny that for future science to ignore the obvious would reduce its legitimacy; yet the point is that the authors score points at the expense of dumb intellectual mistakes that have not even happened. Technical prose always contains less (and thus more) than meets the eye. The untrained reader is led to believe that the analysis has made important inroads.

The following passage is even more striking in the way it derives authority by dressing up its "implications" in pretentious technical terms.

> The empirical results described in the preceding section have a number of implications for macrodynamic theories of crime causation and for ex ante forecasts of future levels of U. S. crime rates.[22]

Only an expert knows that "macrodynamic theories of crime causation" contains both a meaningless adjective and a redundant noun. "Macrodynamic" does little work here; few understand it to have sharp and special meaning. Of course technical experts use terms like "macro" and "micro" all the time, hoping thereby to conceal the emptiness of their work. And the word "causation" is redundant; "crime" would have been enough. Going further, the term "ex ante" is completely unnecessary, as is the adjective "future." One cannot "forecast" the past or present. Finally, "crime rates" do not have "levels." Rates are rates. That is all one needs to say. It is well and good for the writers to spell out the "implications" of their work; but they should do so in ways that have substance. Science's incredibly differentiated discourse is ordinarily called jargon in a positivist culture, thus failing to see that positivism endorses jargon profoundly. This rarefied technical language makes it appear that intellectual work is being accomplished. The analysis claims to bear on "ex ante forecasts." Are there any other kinds of forecasts?

The problem here is not poor writing, as literary positivists would maintain, obscuring the fact that common sense simply means commonplace. Appeals to everyday clarity have no constitutional value; clarity is governed by convention, notably by the level of public intelligence. There are no unambiguous standards of good or bad style in literary presentation; there is only gesture and thus a notion of gestural adequacy. Smart readers can understand "difficult" prose, whereas prosaic prose might fail to speak to them. The issue here is not linguistic or literary felicity. Instead, science dresses up its "results" and "findings" in grandiose terms precisely because they are so empty, duplicating other studies with a penchant for addressing methodological minutae. The volume of technical gestures in science is in inverse relation to their meaning. This is not to say that the concept of meaning ought to be governed by conventionalist criteria; that people "understand" something does not mean that they *understand.* Ambiguity frequently achieves greater clarity than the standardized commonplaces of the time. After all, the positivist philosophy of meaning fails to excavate history from ordinary language.

Some science conclusions are less hortatory or scolding than ecumenical. When little has been learned, one can descend to platitudes, especially ones seeming to resolve intellectual differences.

> Meanwhile, there is much to be learned by both Marxists and non-Marxists about the class structure of gender and more common forms of delinquency.[23]

Of course "there is much to be learned"; science in saying so convinces no one not already convinced. The term "meanwhile" is key here. "Meanwhile" means until the larger issues of theory and method are resolved. Meanwhile...everyday science proliferates on both sides of the ideological line. Until that time when everything is known—namely, never—people can drop their ideological bickering and get down to the task at hand: trivial empirical research. The authors are trying to domesticate Marxism in the name of criminology, an impossible task. Using a little Marxist language, they pretend to be Marxists. But in the end the truth emerges; however irresolveable deep theoretical—political—differences may be, both camps can form a united front around piecemeal, methodologically driven research.

Most science conclusions proposing "further research" do so on unambiguously technical grounds. They do not mince words; their articles are trivial, and whatever insights they afford will be deepened by technical improvements.

> Future studies along these lines may require different methodologies from that employed here...[24]

> [T]he model is presented with the hope that others will use it to generate additional propositions.[25]

> Such processes, however, require theory, measurement, and data analysis beyond the present state of the art.[26]

Whatever reservations journal science has about itself are usually methodological in nature; thus it calls for itself and others to correct them in the future. Once methodology is made to drive science, it is unsurprising that science would call for better methodology. This not only legitimates one's own future work but gives license to the present trivially technical analysis.

An assumption of methodological programmatism is that knowledge evolves continuously. Thus one can talk about "the present state of the art" in presaging future improvements in the base and techniques of knowledge. State-of-the-art is a notion more appropriate to technology than epistemology. While tempting, the implication that knowledge is somehow technically, even theoretically, constrained by its underdevelopment simply apologizes for the choices scientists make in the present. One can excuse one's own work by suggesting that the resolution of its aporias awaits technical and intellectual breakthroughs. But in social science there is nothing inherently blocking those breakthroughs, especially when it comes to what the last passage calls "theory." In fact, the last passage is really saying that the expansion of knowledge awaits certain technical breakthroughs in method; there is no reason why the theory cannot already be improved to deal with present analytical problems.

This technological optimism pervades journal science—indeed, virtually all intellectual activity since the Enlightenment. It is assumed that mysteries gradually dissolve under methodical scrutiny by professional intellectuals, now academics. Yet this assumption is mystery itself when it fails to make good on its own promise. I maintain emphatically that the journal science published in many academic disciplines does not advance knowledge at all but merely repeats itself in professional self-aggrandizement. Just look. Trivia and technical fetishism beget more of the same. In any event, there are no technical limits to reason blocking new insights, methods, data collection. Only technology bumps up against cosmic limits. The intellectual life is unconstrained.

Thus methodological programmatism calling for more and better techniques in the future both exonerates bad science and enriches optimism about inexorable disciplinary advancement. Both are false and both must be opposed. There is no excuse for positivism and there is no guarantee of future enlightenment. Precisely the opposite: Knowledge is constitutive even in its representationality; and the future follows dialectically—thus discontinuously—from the past and present. Programmatism has no excuses for its own lassitude. Nor does it have evidence guaranteeing progress. Instead, regression as a technical concept contributes to civilizational regression.

Finally, a few articles conclude with a different kind of programmatism. Instead of calling for better science, they hint at "policy implications," attempting to programmatize a social and political agenda flowing from their research.

> This research has several policy implications...If school personnel wish to support interracial sociability in desegregated schools, they should try to provide a classroom environment that promotes stable interracial friendship choices.[27]

> Alternatively, and perhaps more critical from a policy standpoint, do they also hold promise for making occupational roles more gender-neutral? Answers to these questions are crucial for assessing the prospects for closing the gender gap in earnings by the end of the century.[28]

At least the concept of "policy" bears more relation to the social world than do endless methodological refinements. Yet the discussion of "policy implications" has become a thoughtlessly routine gesture in its own right. Journal scientists who cluster in social science disciplines sometimes feel the need to justify their research in terms of social and political relevance; hence they end their concluding discussions with a nod to "policy." Seemingly broadening the typical methodological programmatism, this consideration of policy is no more enlightening. It strengthens the view that the social and political relevance of research bears directly on governmental policy—but not on other, perhaps more radical, social interventions.

This assumption is born of welfare-state liberalism addressing social problems piecemeal and then throwing money at them. American social science, if it considers the social implications of research at all, for the most part accepts this welfare-state liberalism; thus social improvements deriving from intellectual insight are to be carried out by the government or state. The very concept of policy locates a policy agenda in the realm of government; the state is held responsible for the amelioration of pressing social problems conceptualized as barriers to overall economic—capitalist—progress. Of course, we could discuss the social and political implications of research in different terms, undoing the close relationship between academic social science and the state. Instead of thinking about the relevance of research to pro-capitalist government, we could conceptualize research's relevance in terms of its facilitation of counterhegemonic resistance to that very state apparatus. These are not just words; once we truncate social and political relevance to policy, we entrench the responsibility of the state to solve social problems. In turn, this fails to notice that it is precisely the state (in its subordinate relationship to a private enterprise economy) that causes many of the social problems in the first place. Policy is just another way of formulating the state's own coping mechanisms with which it addresses the unplanned excesses of the exuberant business growth it facilitates.

Marx said originally that the capitalist state is an "executive committee of the bourgeoisie." Journal science is part of the state apparatus to the extent to which its findings contribute to policy formation and thus discourage other political applications of intellectual insight. Journal science alternatively could apply its results to the improvement of social and economic justice, particularly the elimination of economic inequality, racism, sexism, and the domination of nature. While some journal scientists may privately uphold this extraparliamentary agenda, no one talks about research in these terms. At best, "policy implications" is a euphemism for more sweeping political transformations. Typically, though, journal scientists, as welfare-state liberals, sincerely intend that their research contribute to gradual social betterment via the state. Thus they consider, if ritualistically, the world from a "policy standpoint." Radicals not only oppose the particular policies of capitalist and state-socialist governments but the very concept of policy as an illegitimate truncation—thus neutralization—of political activity.

Just consider the passages above. In the first instance, "policy implications" have to do with "support[ing] interracial sociability in desegregated schools," a profound aim indeed. Journal science, in its elaborate methodological protocol, suggests that this aim can be achieved by providing "a classroom environment...promot[ing] stable interracial friendship choices." It is hard to imagine a more trivial policy agenda, especially when many urban schools are moving further away from integration. And the notion that overall racial inequality will be seriously addressed by leading minority and white children to make "stable interracial friendship choices" reveals the poverty of liberalism today. Journal science moves beyond methodology to the real world only to arrive at profundities like this.

The second passage is denser with innuendo and construction. The authors speculate about the usefulness of their findings for "making occupational roles more gender-neutral," leading, they suggest, to a reduced "gender gap in earnings by the end of the century." First, the political agenda of state policy has very little relevance to what the authors call "occupational roles," except perhaps in certain public-sector agencies. The state has little influence on private sector work arrangements except to support them. And the notion that the state might take responsibility for making work roles more "gender-neutral" seems unlikely given the state's own reluctance to legislate serious changes in the economic situations of women and men. Thus, it is not even clear that the gender neutrality of work roles has anything to do with "policy."

Next, the authors simply assume that making work roles more gender neutral (whatever that might mean) will reduce the income gap between men and women. Yet the income gap stems from an overall sexual division of labor relegating women to the realms of unpaid and underpaid labor. The so-called gender neutrality of work roles has very little to do with the income gap. If anything, it reflects it. Finally, the authors repeat a cliche apparently without thinking about what they are saying. Why should women await "the end of the century" for some redress of their economic inequities? Again, welfare-state liberalism assumes that while change—progress—occurs inexorably, it happens only through an accumulating series of small steps. Like the growth of scientific knowledge, social progress is thought to happen incrementally, each small step extending the one before it. Thus, policy programmatism assumes that history, like journal science, inches forward. Science keeps history on course.

To be accurate, policy programmatism is much rarer than the methodological kind. Few scholars conclude with a glance at the social relevance of their work; writers on gender, race and class do so most frequently. Policy programmatism resembles methodological agenda setting in that it suggests modest transformations. And it writes as if its goals, whether more interracial friendship or better measurements of variables, were unproblematic, indeed shared by other responsible members of the academic community. As always in the deauthorizing text of science, the literary and political interests of the writer are suppressed except for their unexciting desublimation in the "discussion" section of science articles. Not only is it easy to call for better science and policy; it is really to call for no change at all. Tomorrow's articles and policies will flow uninterruptedly from the articles and policies of today.

Chapter 9 Journal Theory

Journal science proliferates tables, equations, graphs, models, pictures, number—figure. It derives from authorizing literatures narrowly circumscribing the analysis. It reports itself representationally, promising the world's virtual presence in the text. As such, I have argued, journal science pens a metaphysical world precisely in order to bring it about. Science, in seeming to reflect laws, would create them. Yet a number of people appear not to do this kind of work. They do not write with empirical data; they do not gesture the data figurally; they seem to reflect nothing—only to reflect. They are called theorists, disciplines' gray eminences who devotedly protect ideas against their invigoration. Other theorists crudely beg empirical verification. My task in this chapter is to consider theory as a peculiar literary practice in journal science; my contention is that theory differs little from the figural work of empirical analysis. Theory claims difference not to vitiate journal science but only to abet it.

There is overwhelming sentiment among midwestern empiricists that theory is a waste of time. Thus very little theory gets published in the mainstream sociology journals. No matter how prominent theory may be in the minds of those who either hate it or do it, it claims little prominence in the journal pages. What I talk about in this chapter is nearly as rare as four-leaf clovers. Theory's voice can scarcely be heard. Yet I (a theorist of sorts) must consider it if only to undo the impression that in publishing theory mainstream science journals provide speculative leavening for those bred entirely on a diet of positivist representation. Indeed, the literary practice of theory differs little from the narrative craft of empirical analysis. Most theorists and empiricists do not see this; they perpetuate the differences because they do not attend to their continuity.

This is not to say that the theory of journal science is all theory; indeed, my version of theory would be rejected out of hand by most. When mainstream sociologists talk about theory, they usually mean speculative writing deriving from or otherwise addressing concerns raised by people considered to be the discipline's founders: Comte, Spencer, Durkheim, Weber, even Marx, later Parsons and Gouldner. Political science's theorists are called political philosophers—Plato, Aristotle, Aquinas, Hobbes, Locke, Rousseau, Mill, Bentham, also Marx, Rawls, perhaps Dahl. For most disciplinary writers theory engages with other vouchsafed theorists. For some, theory is any writing that does not figure itself representationally—so-called qualitative research. But I have shown that this notion of theory is too broad; all sorts of so-called qualitative sociologists narrate their work in exactly the same way as quantitative empiricists, often using figural techniques (including quantitative analysis).

Journal science narrates a certain world. Journal theory only appears to be discontinuous with quantitative and qualitative approaches to figural representation. In fact, it is exactly the same; only its figures are different. Their totemic character is the same. Whereas empirical analyses move from literature review to research design, methodology and the report of findings, journal theory remains at square one, endlessly transmogrifying certain texts sanctioned by the disciplines within which they work. As such, theoretical writing suggests and then defends disciplinary boundaries from those who interrogate the inalienable rights of disciplines to exist—and exist apart. Far from being marginal to mainstream science, theoretical writing selects and omits certain grounding texts from which science narratives may then proceed. After all, journal science proceeds from disciplinary authority; its problems are derived from opening examinations of traditions within which people formulate intellectual problems in certain legitimate ways. Journal science methodologizes these narrow topics; journal theory helps tend the literatures from which methodologization proceeds in the first place.

It seems obvious that disciplines exist. Of course they do; the university is differentiated into departments by discipline: anthropology, mathematics, psychology. Yet their existences must be reproduced through certain literary as well as organizational practices, especially in the newer disciplines. It cannot be assumed that every would-be professional knows that (or how) psychology lays claim to be a specialized discourse. Theory reminds the discipline of its forgotten prehistory. It establishes what it means to do disciplinary work by voicing certain names and traditions; authority is thus authorized. As well, positivist theory entrenches discipline by reducing its own speculativeness to researchable, testable propositions. Thus theory plays a special programmatic role of its own, generating hypotheses for others to address empirically.

In this chapter, I want to address three moments of theoretical presentation in journal science. I begin by discussing the canonizing function of theorizing, naming names with which to orient disciplinarily legitimate work. I then discuss theory's data—its analytical activity. In particular, these activities consist of literature reviews and technical hermeneutic work. Theorists also programmatize and propositionalize theory's implications for subsequent empirical work. In these senses, theoretical articles are very similar to empirical articles; what they lack are only quantitative data. Theory makes way for empirical work and constitutes itself as a form of legitimate analysis in its own right. Vouchsafed theoretical writing is as technical and trivial as the work it prefigures.

Disciplining Discipline: Theory Names Names

One can recognize a theoretical article not simply because it lacks quantitative methods; empirical analyses using qualitative methods (e.g., participant observation, ethnography, archival and historical work) canonize and figure themselves in the typical ways. Theoretical science is distinctive, first and foremost, in that it names the names of disciplinary founders whose original concepts it clarifies and then, sometimes, tests against data.

> In 1937, in *The Structure of Social Action,* Talcott Parsons sketched an initial attempt to construct what he described as a voluntaristic theory of action, extending the model of rationality used by economists and systematizing the historians' conception of purposive action.[1]
>
> This paper presents a new framework for addressing the classical concern with cultural systems as totalities. Its approach is consistent with Durkheim's focus on the relationship between social organization and systems of classification (Durkheim 1915; Douglas 1966; Schwartz 1981). Unlike Sorokin, Marxists, and mass-culture theorists...[2]
>
> In the 1930s Parsons not only coined the phrase "voluntaristic means-end schema" but also "intrinsic means-end schema" and "symbolic means-end schema" (Parsons 1935, pp. 300-305; 1936, p. 87; 1937, pp. 56ff, 133, 141, 109-18, 257-64, 285-88, 383-90, 404, 565-66, 645, 653-58, 673-77, 683-84).[3]

Many theoretical articles in sociology, the journal science I am sampling here, name a small number of founding thinkers as the discipline's prime movers, the people from whom intellectual problematics and a common method are said to derive. Parsons figures most frequently in recent theoretical writing; such people as Durkheim and Weber follow close behind. On one level, these articles can be read as hermeneutic contributions to theoretical scholarship, as I suggest in the next section. On the other hand, one is meant to read them also for the frequencies with which they name certain names—like Parsons. The discipline authorizes itself and thus disciplines people who work within it by constructing a pantheon of originary thinkers who defined the discipline in the first place. Whereas the particular literature reviews beginning empirical analyses mention a variety of names who work closely on a narrow problem, theoretical articles are grandiose literature reviews acknowledging the founders of the discipline per se. The focused literatures surrounding particular research problems are disciplined by the larger literature constituted in theory science. In turn, theoretical practice, as we see below, either clarifies concepts offered by these original thinkers or suggests the empirical implications of those concepts for everyday empirical research.

Theory constitutes disciplines that otherwise have no self-evident boundaries or intellectual problems.[4] Aristotle initiated political science, Freud psychology, Comte sociology—yet except in the case of Comte, who authored the neologism "sociology," none of these people self-consciously styled their work in disciplinary terms. They could not have; disciplines did not exist. The academic division of labor was nonexistent for Aristotle and only rudimentary for Freud and Comte. These people are invented as disciplinary guardians by subsequent people who write theory and thus discipline thought. Aristotle did not theorize; he attempted to constitute an entire intellectual universe, both empirically and conceptually. He theorizes only for those of us who extract theory from him. In naming him a theorist, he authorizes the disciplinary work that some of us do, especially political scientists.

Theory, in constituting a discipline, includes some and excludes others. It continues the work of others who have already rewritten certain original names. Disciplines by now know their founders; the issue for theoretical scholarship is what exactly to make of them. Revisionism revises debates already held; Parsons is the object of struggle between different camps claiming him in their names. As authorizing activity, theorizing names a discipline's legitimate names and then puts a certain interpretive spin on them. The point is not the interpretation; it is the extent of coverage and controversy given to people. The more ink spilled about disciplinary founders, the more constitutive their work becomes for those who do disciplinary work. It is no accident that Parsons dominates sociology; more theorists write about him than about, say, Aristotle or Adorno. One might observe that Parsons already saw himself as a sociologist, having read Durkheim and Weber as disciplinary founders. True. But Parsons's reading of these other figures becomes a central theoretical topic when it has already been decided that Parsons is what it means to be a sociological theorist today.[5] This decision has been reached. The evidence of this is that people continue to write about Parsons as if he invented sociology; whether he did or not is immaterial.

Thus, in naming people, theory adds value to their canonical status. Even in dispute, Parsons's ideas are constitutive for American sociology—today more than ever.

> This analysis permits a tentative resolution of many of the contradictory interpretations that have been advanced in less historical treatments of Parsons and, with this, a better understanding of what his early method was and was not.[6]

One must keep in mind that this passage appears in an article surrounded by highly technical empirical articles. The article did not appear in a journal devoted to sociological theory but in the mainstream organ of empirical sociology. Few working empiricists know, or need to know, anything about Parsons's "early method." Nothing could be further from their workaday concerns—data, sampling, method, statistics. Yet the article adds value to the constitution of Parsons's alleged authorization of sociology as a legitimate discipline; theory in promising to resolve "contradictory interpretations" of Parsons legitimates the discipline thought to rest on the edifice of Parsons (and his reading of the original founders). Whether or not the discipline actually rests on Parsons, in the sense that his work issues in empirical propositions, is irrelevant. Theory, in talking about early Parsons, makes Parsons a sociologist of record and thus defines disciplinary work in his image. Theory, in constructing original thinkers as disciplinary thinkers, provides sanction for the everyday empiricists who work in what they take to be its image. No matter that no one ever debates the meaning of the term sociology or interrogates the discipline's right to exist as a valid and separate intellectual project. Theory reminds people that they are disciplinary workers—thus forestalling interrogations of discipline.

Parsons talk is disciplinary talk when certain generations of professional academics take Parsons to be an originary sociologist. However little that talk contributes to research design and empirical analysis, it provides a broader framework within which everyday research can be carried on. Disciplines could not even publish journals if they did not constitute disciplinary work in the images of founders. Theory disciplines disciplines when it suggests their prehistory and maturation. Disciplinary work is Parsonian here not because everyday empiricists use his concepts but because the discipline has come to depend on its Parsonianness (or Durkheimianness or Weberianness—but not Marxness) for its identity. If Parsons was a sociologist, then to talk about Parsons constitutes disciplinary talk and thus legitimates disciplinary work. From time to time—occasionally—the discipline reminds itself of its disciplineness—its having evolved from the pens of founding thinkers who charted its intellectual course and established its analytical frame of reference. No one but the theorists reads Parsons; yet Parsonian readings are published as disciplinarily constitutional work. Parsons talk reminds us of our common academic culture.

Ironically, Parsons survived his own institutionalization. He was still alive when people began to constitute him as a disciplinary monument.

> In his final publications, Talcott Parsons faulted his commentators for neglecting to realize that he "was 'a child of [his] age' and, in working on *The Structure of Social Action* ..., thought in terms of the intellectual environment of the time" (1978a, p. 1351; see also 1976b, p. 364). What particularly perplexed him was inattention to the methodological implications of his early move into sociology from economics.[7]

In reading others' readings of him (and thus of his readings of even earlier thinkers), Parsons understood the constitutional nature of theoretical interpretation. He understood that he stood for sociology and expressed irritation that commentators did not get him right. Of course, Parsons could have simply corrected them; after all, he was alive to do so. But he understood how the interpretation of his work was constitutional for the emerging discipline of sociology. He was concerned about how his work would inform the discipline; he realized that it was not enough to do new work. Instead, he had to attend to others' archaeology of him.

Simply writing about Parsons makes him important. Not to write about others beyond the ken of disciplinary sociology compounds the issue. Journal theory plumbs Parsons for canonizing effects. More important than his ideas themselves is the attention paid to him (and not to others). Journal theory coexists with journal science as its conscience; it reminds ordinary writers that they have a larger identity with a common core of disciplinary concepts, methods, practices. Parsonianness is less bona fide theory than cultural identity. Theory exhibits and thus reproduces cultural fit—in this case, what it means to share discipline.

> Parsons' point in coining a new term was that this third combination of analytical means and ends of social action is distinctive and also irreducible (e.g., 1935, pp. 298-99; 1937, pp. 79, 81-82, 209-215, 221, 251-61, 486 n2, 645 n1, 660-61).[8]

In talking this way about Parsons, we add value to the discipline he is thought to authorize. The closer the interpretive attention paid to the fine points of his oeuvre, the more everyday empiricists gain in the way of a disciplinary imprimatur. This seems most strange. Yet it is not if we realize that Parsons articles do not directly advance empirical knowledge, and thus the discipline as a whole, but defend discipline against self-doubts. Journal theory reassures journal scientists that they belong to a common world; Parsons himself would understand this in Durkheimian terms—Parsonianness functions as a collectivizing agent.

This is not usually how theory is read by theorists. They passionately invest in interpretive skirmishes. Consider the following long passages.

> To date, few scholars have tried to resolve or even sort out these wide interpretative discrepancies. Indeed, only Alexander (1983) has come near to a systematic effort in this area. Holding that *Structure* articulated an "antipositivist", "anti-empiricist", comprehensive, nonreproductionist and nonrelativistic method, Alexander nevertheless acknowledges certain contrary textual evidence and argues, accordingly, that even as Parsons formulated his methodological program, he occasionally harked back to certain "positivist", "empiricist", and reflectionist notions, thus revealing "methodological ambivalence", "confusion", "tension", and "profound ambiguity" (1983, pp. 10- 13, 45, 94, 152-56, 440 n. 17; cf. Bershady 1973, pp. 86, 111, 153-54; Martel 1971; Menzies 1976, p. 160). The trouble with this kind of argument is that it is asociological...[9]
>
> A substantially similar interpretation has been suggested by Adriaansens (1980), Alexander (1983), Barber (1986), Bershady (1973, 1974), Bourricaud (1977), Lidz (1986), Martel (1971, 1979), Proctor (1978), Rocher (1974), Savage (1981), and Schwanenberg (1971, 1976). According to all these sympathetic commentators, the early Parsons is is the pioneering exponent of the "theory-directed" or "analytical view" of science—a conclusion that ironically echoes the more adverse judgment of critics that Parsons's approach derives more from "high... abstractions" than from empirical findings and historical research (Mills 1959, p. 33; Homans 1962, p. 46; Zaret 1980, p. 1197; see also, Merton 1949; Moore 1958).[10]

There are two significances here. Simply by amassing huge interpretive literatures addressing the meaning of obscure formulations in Parsons, theory articles constitute a Parsons-driven discipline, a discipline within which the name Parsons connotes disciplinary status. Where Parsons appears, there shall one find sociology. Yet the first passage dismisses certain approaches to Parsons interpretation as "asociological," rejecting it as undisciplined without reference to explicit criteria constituting sociological analysis. This is circular; Parsons as a name authorizes disciplinary sociology to disqualify certain interpretations of him as undisciplined. It does not really matter what the substance of these disputes is; their substance is in their form. That mainstream science journals publish theory only when it names certain authorizing names in reconstituting the common code of the discipline is extremely telling. Beyond that, it matters very little to working scientists that Parsons essayed an "analytical view" of science or exhibited methodological "ambivalence." Parsons sociologizes the other work sharing the covers with the occasional theoretical article.

This is necessary when the majority of journal science reads like mathematics. Without being disciplined, everyday journal writing could be mistaken for work in other disciplines, especially when its text is littered with figural formulations leaving little trace of sociological mind. Editorial decisions to publish theory arise from the anxiety that science has become so figural that it threatens to lose its grounding in an original literature; the danger is that the discipline will lose its reason for being. Figural sociology could as easily be figural economics, political science, even biology. The fading presence of the science author requires nuanced theoretical work to exhume science's ground in a disciplinary literature (even if that grounding has been constructed post hoc). The name Parsons occasionally interspersed in journal pages makes way for a science increasingly unreliant on names to do its intellectual work. Theory's text has to be read against the backdrop of everyday science crying out for historico-theological buttressing. Seen in this light, theory becomes hagiographical activity carried out by acolytes.

Theoretical Practice I: Citation, Interpretation, Constitution

The two long passages cited in the previous section are larded with references to other writers. Indeed, theory's text is figured when it is heavily reliant on names, dates and page references. Theory's data, these references constitute the rhetorical substance of journal theorizing. Rarely does a theoretical article come unequipped with a huge bibliography as a context within which these frequent parenthetical references can signify meaning. The names named by journal theory constitute the discipline; in naming names, theory disciplines the empirical work surrounding it. Whereas theoretical articles open by promising to resolve interpretive disputes surrounding disciplinary founders (e.g., Parsons), they spend the bulk of their time citing, interpreting, and thus constituting a disciplinary culture within which positivist representation can then proceed. Theory, in effect, constitutes a disciplinary world within which common intellectual assumptions can be taken for granted; at least, we can assume that we know and revere the same originary names, thus establishing (and reproducing) our professional identities. In belaboring interpretive issues, disciplines differentiate themselves from other disciplines at a time when the discourses of figure and number seem hegemonic.

> One purpose of this paper is to eliminate the concept's vagueness. After reviewing the meanings attributed to voluntaristic action in the literature, the concept's distinct analytical components and meanings are specified in Section II. Parsons' writings are drawn upon directly. But by bringing greater clarity

> to Parsons' original wording (1935, 1937) as well as to his later summaries, comments and elaborations (e.g., 1968b, 1970a, 1974a, 1975a, b, 1976, 1977b, 1978a), the concept is reformulated.[11]

A model of theoretical presentation, this passage promises to "reformulate" a central Parsonian concept. Theory will work with "the literature" as well as with "Parsons' original wording [and] his later summaries, comments and elaborations." All told, this work promises to improve on a crucial Parsonian idea through theoretical labor. At least, that is the way the theory text wants to be read. And it is certainly plausible to take it on its own terms, evaluating its adequacy as interpretive work. Such literary work is not impossible; nor does it have to be unimportant simply because it poses a narrow project for itself. But I want to read theory as a literary auxiliary to everyday science, in effect constituting a common discourse within which disciplined analyses may proceed elsewhere in the science journal beyond prose. If the theory is artistically compelling, it will provoke new debates, even entering the literature itself as an important voice. That is entirely possible.

Yet that is not what is most important about theory's text. Just as science signifies a certain metaphysical and thus political world whose representation it hastens, so theory suggests common disciplinary discourse. Theory is not required to show a close relationship to empirical work even if sometimes it seems to achieve it (see the next section); the important thing about theory is that it adds value to sociological discourse. Theorizing gives history and thus legitimacy to what might otherwise be read to be nameless, even wordless concepts. When theory, in naming the discipline's names, reinforces its common culture of discourse, theory works with those names; citation gives way to interpretation and thus the constitution and reconstitution of the discipline. It also makes way directly for empirical research. Theory goes beyond sheer naming. Citation facilitates interpretation, which in turn adds value to theory's sense of discipline. While it is important to read theory's text simply in the names themselves—Durkheim, Weber, Parsons, for example—theory's overt text is also a text in its own right. Theoretical practice is bound up with the endless interpretation of original texts.

This is essentially a biblical practice; interpretation is thought to disclose faith, thus constituting the faith circularly. To be a sociologist (like a Christian, Jew, or Moslem) is to do sociological witnessing—attending to the interpretation of original sociological texts as an act of fealty. Sociological theorizing is thus to theorize about sociological theorists. While a truism to people doing this work, it is far from obvious who the legitimate theorists are, especially when the founders are only constituted that way post hoc. We make the founders founders by naming them;[12] Weber did not attempt to build a discipline but to comprehend world history in counterpoint to Marx's version of it. We display our commitment to sociology and thus legitimate the discipline through our disciplined interpretive work, giving value to certain thinkers and not others and thus making way for the empirical work people do in the name of sociology.

Theory articles name and then interpret other theorists. But these are not democratic relationships. Positivist theorizing subordinates theory to empirical research; theory constitutes the discourse making this research possible by subordinating itself interpretively to certain thinkers thought to ground sociological analysis. Positivist disciplines embody a crucial irony: They lack correspondence to a pregiven disciplinary object. Sociology constitutes an object for itself and thus gives itself a reason for being only in the self-referential work of theoretical interpretation. In naming and then interpreting Parsons, we convince ourselves that we are sociologists, inhabiting a common universe of discourse evidenced by our affiliation to certain central texts. As Christians expostulate New and Old Testaments, sociologists take their cue from the founders (who become founders only because we find them at the disciplinary beginning).

Theoretical practice is necessarily subordinate to traditions. No one claims to be the new Parsons; people can only improve Parsons, not replace him. Self-pride threatens to undo discipline. Because Parsons exists, we exist as sociologists. We cannot bypass Parsons, Durkheim, and Weber; they give us our identities as sociological theorists and thus as sociologists. Not only does this solve the circularity of disciplinary existence in a deceptive way—inferring sociology from the people we name founding sociologists—but it also disempowers theorists from thinking intellectual problems anew. A "sociological" theory is one corresponding to the extant disciplinary discourse. Other kinds of theory are illegitimate—nonsociology. This is crippling for people who genuinely want to rethink old problems, even undoing the discipline ensnaring them in thoughtlessness. A scolding faculty member at a job interview once told me that my work on Freud was illegitimate because, presumably, it did not "give sociology a chance." I read the wrong mentors; by definition my theory was nonsociological, thus false.

This is not only a temperamental quirk. It is ensured in the very practice of disciplinary theory; to theorize is to mouth and interpret holy words, thus sanctioning iterative work within that discourse. As a result, other discourses are disqualified out of hand. I am not talking here only about other disciplines; they might be wrong too. Discipline itself is the problem when its fixation on method and its authority structure stifle thought. And disciplines modeled on natural science publish themselves in journals; the occasional theoretical, historical, or philosophical article reminds everyday scientists of their disciplinary identity—binding them to a single vouchsafed method and making their work iteratively trivial under the tremendous influence of established disciplinary canons.

One might suppose that theorists would be the freest intellectuals in academic disciplines. A well-known scholar appraising the doctoral program of a department in which I used to work admonished us for being too "nitty-gritty," not paying enough attention to "big questions" of theory. Thus I was hired. His point was that theorists can synthesize the myriad researches of daily science as well as provoke new research. Although not wrong in principle, that does not happen; theorists are no freer than others who work in disciplines and rely on disciplinary sanction to publish their work and thus advance themselves. Contemporary theorists must pay homage to the people named in theory articles as important; the work they are left to do is trivial and demeaning—explication of texts.

Thus theory articles work interpretively on disciplinary founders established circularly in theoretical work itself. Like capital, authority tends to perpetuate itself, even to accumulate in fewer and fewer hands. Once a discipline has constituted itself around founding names, whoever they might be, it is virtually impossible to undo their hold on subsequent thinkers. Theoretical work is inherently duplicative, iterative, interpretive; no one writes new theory, for to do so would challenge disciplinary boundaries and thus disciplinary identity. To be a disciplinary theorist (here a "sociological" theorist) one must engage with the founding names to whom one adds value in one's own interpretive work. Theoretical novelty is inherently disqualified in that the project of theory is defined as interpretive and thus constitutional in nature; theorists keep the archives of disciplines, occasionally dusting them off, revising their translations, and otherwise worshipping hagiographically at their feet.

If one does not do interpretive work in one's theoretical writing, the subject of this section, one may actually derive empirical propositions from one's speculations, so-called theory building. Indeed, if there is a dominant tendency in mainstream sociological theory, it is away from sheer interpretation toward a blend of empirical and conceptual work; theory is increasingly harnessed to the empirical research enterprise, useful only inasmuch as it might suggest testable hypotheses to be pursued in everyday journal science. At the moment in American academic sociology, one finds very few theory articles of any kind. Those that do appear seem to be divided between hagiography (explication of text) and theory construction, the topic for my next section of this chapter. Let me look more closely at theoretical hagiography as a literary style before I discuss empirical theory.

Interpretive theory takes two primary forms. Some theorists read single authors or well-integrated theoretical traditions—Parsons, functionalism, Marx, Marxism. Other theorists attempt to make peace between vying theoretical traditions.[13] Some articles do both. In each instance, the theoretical narrative adds value to a certain conception of disciplinary work simply by naming some people and omitting others; then, as I have said, the theorist subordinates himself or herself to that tradition, precluded from heterodox formulations of discipline issuing in entirely different approaches to the empirical world. Like empirical work, theory as narrative attempts to conceal its narrative nature. Instead, in its own way, it wants to be read as science, albeit methodical explication if not data analysis.

Bibliography length in journal science seems to relate inversely to the extent of an article's figural display; the more figured and numbered the work, the shorter the bibliography can be. Of course, science articles proceed from literature reviews of their own, and this work must be cited in closing bibliographies. Yet theory bibliographies are much longer; in effect, they replace the data and methodology discussions found in science articles. The extensive citation of scholarly work both authorizes one's argument and then introduces a degree of technical density into the prose replacing the density of figural and numerical work. Lacking empirical data that can be reconstituted figurally, theory articles secure the technical appearance of scientificity in their elaborate use and citation of sources. Indeed, the first page of theory articles often concludes with extensive acknowledgments of well-known scholars and institutions making the analysis possible.

> ...In addition to ASR reviewers, several colleagues have contributed comments and criticisms on this paper. Robert Antonio of University of Kansas provided detailed written comments on an early draft. A second draft was presented both at University of Chicago and at a local theory group in Washington, D. C. In Chicago, helpful comments came from William Wilson, James Coleman, Donald Levine, Terry Clark, Wendy Griswold and Thomas DiPrete. In Washington, useful criticisms came from Charles Herrick, Gerald Mara, Keith Breclaw, Terry Russell and Jere Cohen. The final draft was closely reviewed in Washington, D. C. by David Jacobs of University of Oregon, and in Florence, Italy at the European University Institute by Anna Wessely of Eotvos University, Budapest, Hungary. Throughout these rewrites, I have been influenced generally by discussions with Jeffrey Alexander, Dean Gerstein, Cynthia Rutkowski, Ruth Wallace, William D'Antonio and Sharon Rogers.[14]

After reading this, can one fail to realize that the author is well connected and worldly? Shuttling to and from Chicago and Washington, he has garnered advice from some of the biggest names in the discipline. And he has been privy to "discussions" with a host of notables. An acknowledgment this extensive could easily introduce a book; here it introduces only a few journal pages. Theorists have to go the extra mile in authorizing their own work, both compiling extensive bibliographies and citing that work parenthetically in the text and acknowledging the "comments and criticisms" of significant disciplinary citizens. Do the people acknowledged really care? Does the author send them flowers too? Here, subtext is so close to the surface that we do not elaborate deconstructive methodology to prise it free; the writer adds value to his work by showing his connections to a professional community of people willing to read and appraise his work. This may be fine when one has written a long book taking up much of one's professional life; yet this is only an article. What will the author do when he writes a book?

Theory tries to create the impression of its indebtedness; in naming, it drops names. Virtually every theory article contains dense passages thick with citation and acknowledgment, such as this one.

> "One of the most distressing things about the elaborate secondary discussion...is [its] failure to appreciate the importance of economics in the whole picture"—when "the problem of the relations between economic theory [and] sociological theory" was "the focal problem" of *Structure* (1979 letter to Alexander, cited in Levine 1980, p. lvi; 1974b, p. 216; 1975, p. 669; 1981, p. 193). Parsons's recall was not always infallible. These particular recollections, however, are corroborated by a lifetime of statements to the same effect (e.g., 1937b, pp. 454-55, 728; 1959a, p. 620; 1968, p. vi; 1970, p. 828); by the circumstance of his heavy involvement, while at work on *Structure,* in the debates that were then occurring in American economics (see 1932; 1934b; 1935a; 1935b; 1935c; 1936a; 1936b; 1936c; 1937a); and by the fact that, apart from Alfred North Whitehead and L. J. Henderson, virtually all of his early ties and obligations in the methodological area were to economists at Harvard, most notably Frank Taussig, Overton Taylor, and Joseph Schumpeter (see 1937b, pp. xxii-xxiii; 1959b, pp. 5-7; 1968, pp. vii-viii; 1970, pp. 827- 30, 873-74; 1971, pp. 233-34, and n. 11 below). It would appear, therefore, that methodological developments in economics in the years preceding *Structure* formed a context of some consequence for the early Parsons.[15]

This passage remains largely within Parsons's oeuvre, at least in its parenthetical work. The author shows intimate familiarity with Parsons's work and breaks up his narrative with parenthetical constructions ruled by figure—here dates of publication. Where science articles display data in tables, equations, graphs, models, and pictures, theory figures tradition, reducing it to the parenthetical citations of years and page numbers. It is also interesting to note that whereas journal science largely dispenses with specific page references to cited work in its literature-review section, journal theory proliferates page numbers as a way of adding technical and figural detail to its dataless text. Indeed, the dates of publication and pages of cited work constitute a database in their own right.

In the passages below, taken from notes, others are cited.

> At least three meanings for the concept may be found in Alexander's four volumes (as well as his 1978 article): (a) actors' internationalized "normative control" (e.g. 1982a, pp. 81, 84, 96-112; 1983a, pp. 28-29, 58-59, 102, 113; 1983b, pp. 24-25, 35-44, 119-150); (b) actors' "voluntary" responsibility or choice (e.g. 1982a, pp. 87, 96-98; 1983a, pp. 98-99, 107-108, 112-14; 1983b, pp. 25-26, 120-27, 214); and (c) actors' "motivation" (e.g. 1982a, pp. 23-24; 1983a, pp. 39-40, 96, 107, 116; 1983b, p. 31). For Alexander (1978), "formal voluntarism" is actors' inherent capacity to volunteer to act in common, and "substantive voluntarism" is the empirical realization of that capacity.[16]

> It is interesting to note the many commentators who ignore the problem of defining or systematically discussing voluntarism, e.g. Faris 1953; Swanson 1953; Bredemeier 1955; Cancian 1960; Wrong 1961: Bershady 1973; Rocher 1975; Mitchell 1967; Barry 1970; O'Neill 1972; Madge 1964, chap. 4; Kaplan 1968; Burger 1977; cf. Warner's approach to the problem by 1978, and Wiley's expression of frustration by 1979.[17]

All these passages in naming people constitute a mainstream discipline. These people are thought to exemplify good sociology; thus the discipline reproduces itself in its image of their image. Names become a narrative; they are to be interpreted, refined or synthesized. These abundant citations suggest clarification or improvement of vouchsafed concepts. This is legitimate theoretical work in its own right, giving theorists a rationale behind which they can add value to the disciplinary canon and thus protect disciplinary boundaries. Interpretive theory promises either to clarify and improve concepts or to arrive at syntheses of concepts, a subject I take up shortly. Theory narrates the interpretation of "theoretical" work by promising either rigorous explication or conceptual refinement, unburying meaning in an interpretively legitimate way.

This long passage, from notes, is an excellent example of interpretive clarification. Its highly technical nature is immediately apparent; one would have to know Parsons well to make any sense of the argument.

> The evolution of Parsons' choice of terms in eventually arriving at the "fiduciary subsystem" and also at "collegial formations" may be read as a convenient case study of his general development as a theorist, and in particular how his theorizing eventually took a "procedural turn" and thereby acquired a "critical edge." By exploring what Parsons meant by "procedural institutionalization" in the late 1970s, and especially the importance he placed on the interpretation function of law as opposed to its enforcement function, Parsons' thinking may be traced by exploring how he changed his terminology in attempting to convey his understanding at each point. As examples, Parsons separated the problem of enforcement of sanctions from the problem of value conflict by 1939 (1939, p. 47), and by 1956 he separated the former also from both both bargaining framed by rules (Parsons and Smelser 1956, pp. 104-105) and actors' internalization of shared norms (Parsons 1956). Only a schematic overview of Parsons' uses of "fiduciary" and "collegial" can be provided here. On the term "fiduciary", Parsons first referred to "fiduciary influence" in his 1963 essay on influence (1963b, pp. 421-422), but it offers no clear connection to his later formulations. He refers to influence oriented to the interpretation of norms in the same essay, but he does not yet term it fiduciary. By 1966 Parsons talks of "interpretive responsibility" (1966a, pp. 321-22) and of the "associative subsystem" (1966a, pp. 340ff). By 1969 he finally uses the term "fiduciary" in the way employed through the rest of his works (1969a, pp. 508ff; cf. 1971, p. 26; 1975c, pp. 387-88; 1977c, pp. 35ff; Parsons and Platt 1973, pp. 8, 18ff, 45ff, 125ff, 142ff).[18]

It is difficult to overcome the unrelieved monotony of the narrative—no paragraph breaks, numerous parenthetical citations obsessively interrupting the flow of prose, meager payoff by the end. This talmudic scholarship is typical of interpretive approaches to theoretical narration. By the end, the reader is exhausted: Parsons has been further canonized and yet another log has been added to the hermeneutic fire.

Of course, from the perspective of hindsight, this hagiographical compulsiveness is no more trivial or fruitless than the minute mechanical forays of mainstream methodology. The more we consider the narratives of theory and science side by side, the more we see similarities. Both are technical to a fault; neither raises important problems; each is subordinated to prior work and yet spawns further trivial work in its image. After all, both journal theory and journal science engage with their topics interpretively: Whereas theory clarifies dead prose, science extracts meaning from the otherworldliness of figured data. Both narratives pose and seem to resolve puzzles, avoiding other sorts of constitutive work – failing to ask the question to which they are meant to be answers.

Whereas journal science strikes a representational posture to the phenomenal world, figured in data, journal theory aims to reflect a theoretical world; its phenomena are canonical texts, if not what Durkheim called "social facts." In this sense, theoretical interpretation is a positivist exercise; at least, its narrative style trades heavily on a representational notion of textual explication. It is assumed that interpretation can illuminate important published works, clarifying, improving and sometimes synthesizing their insights. Positivist theory seeks to reflect an unambiguous theoretical world through rigorous interpretive and interpolative work. As with empirical research, the aim is simply clarification – bringing to light; it is achieved through the highly figured work of parenthetical citation interspersed into thick discussion.

Theory papers sometimes narrate themselves in terms of prosaically interpretive aims. Other times, they claim more in the way of synthesis. Let me offer a typical example of the first goal.

> This paper outlines the principal areas of disagreement over Parsons's early methodological views and shows how these views may be coherently elucidated by examining the sociointellectual context in which Parsons worked and the social position that he occupied within that context.[19]

Here, the aim is not even as grandiose as synthesis; the paper aims merely to "elucidate" "areas of disagreement over Parsons's early methodological views." While it is fine for people to plumb other thinkers in terms of their intellectual development, it is clear virtually to everyone working in mainstream sociology that Parsons's early methodological perspectives have absolutely no relevance to everyday research. They merely function iconically in the theory text to reconstitute a discipline that knows itself as a Parsonian one – sociology. The theory paper just cited works hard to illuminate the young Parsons, returning to his "undergraduate education."

> The young Talcott Parsons was intimately involved in the context just examined. In the course of his undergraduate education at Amherst under Hamilton and Ayres, he was immersed in the writings of the institutionalists, which then provided his "jumping off place" into the social sciences (Parsons 1959b, p. 4).[20]

Journal theory even understands itself as puzzle-solving, Levi-Strauss's *bricolage,* creating meaning out of fragments.

> With this, all the pieces fell into place. For what Parsons had in effect done was to extract from the midst of the dispute in economics the justification he needed to put sociology securely on the academic map.[21]

The theorist understands his own authorial motives very clearly; he wants to trace the prehistory of a discipline in Parsons's shift from economics to sociology. Theory here helps constitute the discipline in the name of someone taken to epitomize the sociological enterprise by most working professionals. Parsons's own biography figures centrally in this story of how sociology emerged as a credible academic discipline in its own right.

> Parsons accordingly "settl[ed] down at Harvard...with extremely uncertain expectations for the future" and, after staying several years, became convinced that he "could not [count] on a future in economics at Harvard," a realization that, when combined with his developing substantive interests, led him to transfer in the early 1930s to the newly created, tiny, and intellectually ill-defined Department of Sociology (1970, p. 834; 1981, p. 184).[22]

A different theory writer reconstructs Parsons' import for sociology in much the same terms.

> The three meanings of voluntaristic action open the way to reconsidering later developments in Parsons' theory construction. By the mid-1950s, for instance, he distinguished between (a) "normative *motivations,*" or actors' *internalization* of *substantive* norms and (b) "normative *orientations*", or actors' *recognition* of the integrity of *procedural* norms (Parsons and Smelser 1956; by as early as 1937 and 1939 Parsons had distinguished between motivational and institutional "normative patterns"). By the early 1960s, and eventually under the direct influence of Harvard legal theorist Lon Fuller (see note 9), Parsons emphasized the importance of what he called "procedural institutions" in maintaining the integrity of qualities of social life that may restrain or mediate authoritarian pressures of purposive-rationalization (e.g. 1962, 1968a, 1969a, 1971, 1977c).[23]

That same article, though, promises more than interpretive clarification; it also uses Parsons in a synthetic theoretical pastiche, employing him centrally in a new formulation of the theoretico-methodological perspective underlying his work. The following passage is typical of theoretical narrative aiming beyond interpretation toward conceptual synthesis itself.

> The result, in Section V, is a new conceptual framework for comparative political sociology called societal constitutionalism, and orienting hypotheses which begin to move this framework to specific problem areas in comparative research.[24]

In promising a "new conceptual framework," the author seems to avoid the risk of simply packaging old wine in new bottles. In fact, this vaunted "societal constitutionalism" reads very much like the original Parsons (albeit now spruced up with obscure formulations from Lon Fuller and Jurgen Habermas). Interpretation here issues in "orienting hypotheses"—testable propositions— via a "new conceptual framework." A theme I take up below, theory sometimes operationalizes itself directly in harness to the methodological requirements of the empirical research enterprise. Here it is important to note that journal theory uses interpretation for theory construction—the formulation of a new conceptual apparatus with which to derive testable empirical hypotheses.

Where straightforward interpretation gives way to synthetic theorizing, the aim is usually ideological meltdown, a classic literary technique of liberalism gradually edging sharp differences toward a conciliatory middle position.

> Indeed, societal constitutionalism suggests that the left-right dichotomy, as well as any immediate concentration by comparativists on the class composition of different regimes and social orders, should be counterbalanced by a preliminary examination of the extent of integration or control within a society.[25]

Here, theoretical synthesis blunts "the left-right dichotomy." Science in the name of synthesis banishes ideology, ever the aim of positivism.

> If future studies of urbanization and underdevelopment are going to be useful, then they must transcend current theoretical and ideological particularism.[26]

Theoretical synthesis deftly disqualifies political perspectives—especially Marxism—on epistemological grounds. If one founding theorist is good, two or more should be better. Synthesis is sometimes defended in the name of eclecticism, which is itself unexplicated.

> The first step in the proposed theoretical strategy is to extract the key motivating dynamics from these highly eclectic approaches which, as much as any others in sociology, have carried forward the early theoretical tradition.[27]

Theoretical pluralism is really political pluralism; these are code words for middle-class, middle-brow liberalism—anything but Marxism and radical or socialist feminism. Pluralism both disqualifies radical perspectives out of hand—particularly as nonscience—and then constructs the margins as pale versions of themselves participating cheerfully in the scientific accumulation process. Disciplinary Marxism and feminism are imposters; not only do they not speak authentically for themselves but they are made to mouth words that are not theirs.

> There are...serious limitations to the Marxian critique of mass culture, not the least of which is its functional reductionism. The underlying theoretical assumption of this approach, whose untenability I discuss below, implies that structural or institutional practices are automatically transformed into deep-level psychological ones through the agency of media control. According to hegemonists, consciousness is either "false" (i.e., the masses perceive illusion and not reality) or "contradictory" (the masses are confused and their judgment is fragmented). Basically, this is a very simplistic view of humanity and the nature of cultural expression in everyday life.[28]

It is one thing to oppose certain Marxist approaches within social science. Twenty years ago, it was perhaps even credible to talk about "the Marxian critique of mass culture" as if there was only one. But to conclude a science passage with the sentiment that "Basically, this is a very simplistic view of humanity" is quite another thing. By science's own rigorous criteria of representationality, that sort of sentiment belongs on the op-ed page, not in academic journals. One rarely finds the word "basically" as a sign of summary, thus dismissal, in journal science. And to characterize something as "a simplistic view of humanity" without argument or evidence is outlandish with regard to the literary conventions of everyday science. The author clearly cannot abide the egregious simplifications of Marxism, just as the author above rejects "ideological particularism," another formulation of the word Marxism.

The epistemological ideal of synthesis is deeply imbedded in journal science. Theory is responsible for effecting syntheses, thus avoiding "ideological particularism" and making way for useful science. In the passage that follows, synthesis is defended epistemologically on grounds of "robust[ness]."

> Despite each theory's effort at synthesis, none alone completely captures the full dynamics of motivation. It is appropriate, therefore, to attempt a provisional coupling of these five models in order to develop a more robust sociological theory of motivation.[29]

Is robustness meant to call up images of strong masculinity? Journal science lapses into tough talk in the midst of routinely theoretical discussions; although a technical term known to methodological insiders, "robust" adds value to a theory article read by people for whom "robust" could as well describe the flavor of beer.

Journal theory in this way suppresses the notion that synthesis may not be a good thing after all, or even impossible. Epistemological pluralism disqualifies intellectual extremism on unexplicated ontological grounds—for lacking robustness and the like. But unpopular views may indeed be highly nuanced and strongly phrased—nonsimplistic, robust, irreducible to average sentiment. Synthesis is really intellectual and political centering, marginalizing dissent and then shrewdly translating it into mainstream wisdom—methodological techniques, insights, caveats.

> Although the concept of class is central to Marxian and conflict theories of crime (Colvin and Pauly 1983; Spitzer 1975; Greenberg 1977; Chambliss and Seidman 1971; Quinney 1977; Taylor, Walton and Young 1973), this concept has never before been operationalized in self-report research (cf. Hagan and Albonetti 1982).[30]

This is prototypical journal theory; it reviews literature, claims heterodox intellectual traditions and then turns them in the direction of mainstream science—operationalization. The problems here are manifold. The authors assume that "Marxian...theories of crime" exist. Perhaps they do. But Marx did not make crime a central topic. Later social scientists disciplined Marxism by using such terms as "class" and "conflict" in their own research, either implying or claiming filiation to original Marxism. In any event, that there even can be a Marxian theory of crime requires elaborate argument; I suspect that the ultimate Marxist criminology would say simply that capital is theft. But the deepest problem of all is that Marxism is reduced to trivial methodological terms; the authors propose to "operationalize" the concept of class "in self-report research." They do not tell us why but merely assume we know that any methodological and conceptual mainstreaming of Marxism is valuable in its own right.

But instead of fertilizing mainstream journal science, this only coopts and defuses Marxism. The concept of "class" does not make an analysis

Marxist, especially when Weber, too, used the term. The methodologization of allegedly Marxist concepts only reduces a rich theory of history and political practice to the trivially figural terms of journal science. As such, Marxism supplies only another variable in the equations specified and estimated in the methodology sections of typical articles. Proponents of this say that they thus legitimize Marxism in the mainstream; I respond by saying that what is being legitimated is not Marxism at all—a full-blown theory and practice—but only a Marxist name, a Marxist (and, ironically, Weberian) slogan: class.

At root, journal science employs theory to smooth out intellectual (and thus political) contradictions. The linear liberalism of positivism rejects representations admitting of murkiness, ambiguity, especially self-contradiction. A Marxist would respond that dialectical logic and thus dialectical discourse must admit contradiction when the world itself is riddled with contradictions.[31] Indeed, contradictions are only compounded by discourse pretending they do not exist. Positivism as ontology eliminates intellectual difference in order to reduce the differentiation— contradictoriness—of the world at large. Journal science claiming heterodox intellectual traditions, whether Marxism, feminism, or phenomenology, silences them. It vitiates their dialectical representations of the world and thus defuses the dialectic of history itself. Marx saw clearly that the falling rate of intelligence delays the apocalyptic effects of the falling rate of profit—later Marxists add, perhaps indefinitely. Positivist representation, in claiming to achieve synthesis, helps constitute a synthetic world, achieving Mill's open marketplace of ideas at the level of epistemology and even methodology.

Theoretical Practice II: Empiricizing Theory

For many journal scientists, theory is no more than an acculturating course taught to graduate students in order to remind them of their discipline's prehistory and boundaries. For the hardier (perhaps older) scientists, theoretical writing clarifies concepts...

> This analysis permits a tentative resolution of many of the contradictory interpretations that have been advanced in less historical treatments of Parsons and, with this, a better understanding of what his early method was and was not.[32]

...or merely suggests researchable propositions, nothing more. Thus, theoretical writing could as well be found in the methodology sections of science articles, where causal models are specified, speculated. For these scholars, theory becomes theory construction, assembled out of and helping to assemble the pastiche of hypotheses and findings occupying the daily literary work of most science academics.

Theory construction is databased speculation about narrow causal relationships. Eschewing the heights of speculative abstraction, it has been called "grounded"[33] or "middle-range"[34] theory. Many major social science departments do not employ speculative theorists (or even people who can teach the history of speculative theory); instead, they contain one or two people who engage in theory construction, empiricizing and thus orchestrating lower-level concepts forming the core of hypothetical writing in journal science. Theory is useful only insofar as it can be used in empirical research; it is testable.

> What do these findings tell us about the study of urbanization and underdevelopment in the world system? More than anything else, they underscore the importance of testing various theories of urbanization and underdevelopment. This paper showed that two theories help explain the causes of relative urbanization, and three theories help explain the relationship between relative urbanization and economic growth. Failure to test all three views simultaneously would have resulted in incomplete and even spurious findings.[35]

> The findings with respect to the causes of economic growth provide some support for all three theories.[36]

Here, the authors talk about theory not in the sweeping sense of Durkheim's, Weber's, and Parsons's global reconstructions of social structure; rather, theory is a local explanation of causal relationships, nothing more. While it is difficult to check Durkheim's theory of collective consciousness against data, it is far more feasible to operationalize a few of his key concepts and then insert them into causal models specifying their hypothesized relationships.[37] That is typically how journal science makes use of what it calls theory; theory becomes only the prose encoding hypotheses. As such, theory is the literary practice of formulating cause-effect relationships, nothing more.

Against this mundane conception of theorizing, the work of the founders seems monumental, if ethereal, in its sweeping and abstract nature. High theory functions only as canon; theoretical works are museum pieces symbolizing a discipline's birth throes. Lower theory enters ordinary research as the prose surrounding specified relationships among variables. One can talk of "testing various theories," not haggling over deep metaphysical constructions. While disciplinary founders are important as icons, they add little to the methodologization of their ideas. Modern theory construction can remedy that with a dose of technical rigor.

> Talcott Parsons never established a connection between the concept of voluntaristic action and empirical research in comparative political sociology.[38]

This is not to say that one could not establish such a "connection." In fact, much interpretive work in theory is really methodologization—reducing grand speculative constructs to empirically workable size and shape. The writer who lamented Parsons's insufficient empirical grounding draws from him conceptually in order to develop a construction—"societal constitutionalism"—better able to address the nitty-gritty concerns of everyday research.

> Societal constitutionalism orients researchers to view collegial formations as genuine bases of actors' nonauthoritarian social integration even in societies where liberal-democratic institutions and practices are not in evidence. Given this orientation, studies of the relationship between nascent collegial bodies and broader opposition groups in Eastern bloc countries may result in predictions of particular changes in political institutions or of increased levels of social turmoil (that is, empirical predictions, rather than more general references to a "new class" bearing a common substantive interest, e.g. Konrad and Szelenyi 1979; Gouldner 1979).[39]

Again, this is a matter of old wine in new bottles. The theory writer enhances his own career by publishing "theory" in mainstream journals; to publish, the writer must establish the empirical fruitfulness of dead theories. Thus he or she spruces them up, giving them new names and bonding them to the work of cognate authors. Although purely interpretive theory occasionally is published in mainstream journals, empiricizing theory is a better bet, especially when it authorizes itself on the grounds of putative synthesis. The product of this empiricizing work is meager: "[The theory] may result in predictions of..." Predictions are not knowledge; they can be checked empirically, either confirmed or disconfirmed. While deconstructive overdetermination may tend to pack too much significance into single words or phrases, I believe that the small word "may," above, is telling. It indicates the tentativeness of the work accomplished by an obsessively close reading of Parsons (via Fuller and Habermas). All the dense figural work of parenthetical citation bears pitiful fruit—"may result in predictions." One could have made the predictions without this elaborate theoretical edifice, pretentiously named "societal constitutionalism." While neologisms can sometimes do strikingly evocative work, the term societal constitutionalism lends little insight to the empirical work suggested above. The connection exists only in the author's mind.

Theory constructors always assume that the goal of concept formation is empirical research; at least, theory might point the way.

> Each of these definitions may be sharpened and brought closer to detailed empirical research, given the analytical distinctions discussed above.[40]

Theory is thus reduced into the narrowly trivial terms of everyday journal science. Of course, this is the narrative stuff of science articles' opening literature reviews, parenthetically citing authors and their publication dates (but not specific pages in them). High theory—for example, Parsons readings—stoops to the level of ordinary journal science where it merely abets "detailed empirical research." About 99 percent of scientists view theory in this way; it is only as good as the—empirical—work it does. As theory is banalized into theory construction, its differentiation from empirical analysis begins to fade. That is why it is increasingly difficult to distinguish between theory articles and science articles; the theory articles increasingly read like science. They reduce themselves to testable hypotheses or propositions, figuring themselves with numerous parenthetical citations and even occasionally causal diagrams.

The very distinction between theory and science is science's own. The distinction hierarchizes in science's favor. Science claims methodological rigor and empirical relevance, constructing theory as their absence. Theory is distinguished not by an inherently different logic of presentation but merely by its abstract speculativeness. Science has data; theory has only a shelf full of books and articles. Theory can earn its spurs and thus claim limited journal space only by empiricizing itself, turning its narrative away from interpretive work and toward what is called theory construction—the essay of testable propositions anchored in existing empirical literatures.[41] Given the way science constructs theory as speculatively mushy, it is unsurprising that theory clamors to do work judged useful by science. It "sharpen[s]...definitions" and brings concepts "closer to detailed empirical research." The best theory is the least comfortable with its speculativeness; it drives to become empirical analysis, narrating itself in the same terms as science articles.

Some theory articles even have "discussion" sections. Many rely heavily on figure—parenthetical citation and causal diagrams. From the outside, it is hard to know exactly where science ends and theory begins. In this chapter I have assumed unproblematically that one can identify journal theory. In fact, I do this in conventional terms in order to make a point. I read as theory what science calls theory—articles on Parsons, in particular. I deny the science-theory distinction altogether; instead I read them as similar narrative practices, authorizing literatures and programmatizing future research. Science-theory is a continuum. They trivially differ only in that science more elaborately presents itself in methodological terms, displaying its database, sampling strategy and statistical techniques. But this is a small difference when both accounts are driven by representational criteria of faithfulness to the objectively given. Theory through Parsons would assist empirical analysis in its positivist project. As such, there is no such thing as theory in an age when speculation is devalued ("armchair" activity) in favor of high-tech figural expression. Narrative numbers itself both in theory and science; the so-called theory articles are studded with dates and pages, thick with the appearance of technical virtuosity. Whereas we assume that methodological finesse evidences a high degree of academic professionalization, the hagiographic cultivation of Parsons is a kind of discipline in its own right. One has to read the books and closely follow burgeoning secondary literatures. That is largely why theory articles end with huge bibliographies; absent data, they construct their data out of the books and articles lining their office bookshelves.

Theory makes data—citations, interpretations, syntheses—and thus becomes science. Theory is simply science's name for what is yet immature, insufficiently methodologized. Disciplines need theorists as curators; gradually, though, this custodial activity inches toward the prevailing conception of disciplinarily legitimate activity—under positivism, empirical analysis. In the meantime, disciplines trumpet their ecumenism to the uninitiated; by appearing to publish the occasional "theory" article, they suggest more methodological, hence metaphysical, heterogeneity than they really possess. Everyday journal readers actually believe that they are reading theory when they flip casually through the pages of articles on Parsons and theory construction. In fact, they are seeing only the discipline's version of theory—empirical analysis minus quantitative methodology. Everything else is present: canon, figure, empirical programmatism. Science is theory's name for what it would do in a different life. Speculative reason is eclipsed under the inexorable logic of methodological advance—less prose per figure.

Sometimes what appears to be theory, lacking data, vitiates its theoreticity shamelessly. Theory construction differs from interpretive theory in the baldness of its methodological programmatism. No punches are pulled; research is in the saddle.

> To carry out analysis of systemic action with quantitative data, however, requires a formal theoretical model that relates individual actions to systemic functioning. Some research of this sort has been carried out on community functioning (see Marsden 1981; Marsden and Laumann 1977; Pappi and Kappelhoff 1984); but little more than beginnings have been made.

> Here, it can reasonably be said, is where a most promising research-and-theory frontier lies for sociology.[42]

Here, theory does not promise future research avenues. Instead, theory, "a formal theoretical model," is necessary "to carry out analysis of systematic action with quantitative data." It is assumed that theory is a tool, nothing more; as such, theory is virtually equivalent to research. In the first sentence the author recommends theory; in the second sentence the writer refers to the object of the first sentence—"a formal theoretical model"—as "research," some of which has already "been carried out." Deconstruction needs little energy to reveal the driving subtext as text itself. He then turns to a discussion of the deficiences of theory, what he calls "theory building" in the standard jargon.

> One of the most serious defects of a program of theory building and research... is data inadequacy: At a macrosocial level, there is ordinarily too little variation... to test the relation empirically.[43]

Theoretical lacunae will be breached by adequate data, enabling one to "test...relation[s] empirically." The author's serial list of "theory building" and "research" in the first sentence suggests their differentiation. But for all practical purposes they are the same; theory building is simply the literary work of model specification, what above he called the creation of "formal theoretical model[s]." This "theory" writer does not go back to Parsons for his answers; methodology will clarify theoretical muddles when theory has been reduced essentially to modeling work. As the author's narrative unfolds, he drops any reference to theory; instead the argument figures itself into technical discussion.

> To carry out an analysis of the type-3 micro-to-macro relation involves specifying the kind of interdependence that may exist, modeling it (more precisely than I have done here), and then setting up an analysis that tests this model. In this case, it is possible to begin such an examination by quantitative analysis of individual behavior of a sort that is familiar to the sociologist.[44]

The word "modeling" describes the narrative work of "theory building." Speculation is reduced to the formal specification of possible causal relationships, exactly the work of journal science in its opening methodology section. Theory disappears. The term "theory building" promises the demolition of the theoretical edifice even in the limited sense of textual explication and conceptual synthesis. If the Parsons articles are shamelessly trivial, at least their prosefulness to some extent resisted their internal self-figuring tendencies (parenthetical citation, extravagant bibliographies, hermeneutic nit-picking). Theory building is not the leap from quantity to quality: Parsons articles, in their old-fashioned interpretiveness, give way entirely to formal theoretical modeling. I am not saying that theoretical interpretation and synthesis represented a golden age; far from it. Real theory ceased with the old-European philosophical dinosaurs: Hegel, Kant, Marx, Lukacs, the Frankfurt School, Sartre. And theoretical discourse within the social sciences has marginalized these people into philosophy and literature departments where they have also been methodologized and disciplined, not read. Theory is a literary practice refusing to suppress authorial presence in the name of phony methodological rigor. Theory celebrates authoriality and thus opens itself to other voices as a constitutive practice of the good community. Theory is not a prosthetic arm of science nor simply a body of received wisdom. It is narrative that listens to itself write and then makes that self-consciousness thematic; as such, it opposes closure, linearity, monologue, all topics of my concluding chapter.

Theory today has itself become professionalized and disciplinary. Literary and social theorists write incomprehensibly because they fear the incursion of the world into technical and figural display. Whereas science freezes the world metaphysically, theory cowers in fear of the presence of history. Avoiding the covert politics of science, theory eschews politics altogether. As method, theory ironically imitates the methodological machine it so hates and fears. As I explore in my concluding chapter, this is no alternative; theory still loses when it retreats to English and philosophy departments as method. What is missing is public discourse. Science suppresses it in the big chill of representationality; figure suggests *and thus brings about* a frozen, intractable world. Yet the methodologization of theory does not restore thought to writing; it merely defends nonscience as a legitimate expressive, thus personal, strategy. But theory prematurely allows a narrow version of science—representationality—to claim scientificity. One can address, and thus hope to change, the world without figuring—hence, freezing—it. Dialectical writing (and thus dialectical community) is possible even if we are instantly alert to the tendencies of the word "dialectical" to become yet another cliche. This is a dismal alternative: the figural pyrotechnics of metaphysical/political science "or" the sloganeering of theory ("theoretical specificity," "surplus value," "colonization of the lifeworld"—even my own "fast capitalism").

The notion of formal theory construction characterizes the literary styles of journal theory. The casual use of the term "formal" trades on its technical usage implying mathematization in causal modeling.

> Despite the abundance of implicit theories, few sociologists have constructed formal conceptualizations of motivation.[45]

This writer does not specify a model of motivation as an equation to be estimated but merely figures a possible model using graphic representation. While this is legitimate as a heuristic gesture, the word formal does double duty. It implies both model specification, the stuff of high science, and systematic deductive reasoning. Here, the term "formal" adds value to an analysis that remains purely theoretical—that is, nondatabased. Formal theory is shorthand for theory programmatically relevant to subsequent empirical research.

Journal theory's degeneration into a narrow version of science is complete in theory building. The hegemony of methodology over reason eclipses what the Greeks called *theoria*—the ability of thought to constitute a world. Yet, surreptiously, positivism is the more constitutive for all that. When thought is reduced to a research protocol it succumbs to representationality, ensuring that the findings sought will be found—a lawful world. Theory's narrative becomes virtually identical with science when it formulates itself as theory construction into hypotheses or propositions. The current editor of the leading journal in my own discipline, sociology, is known for asking "theory" writers to "propositionalize" their arguments. Mere prose is archaic; the words must programmatize empirical research, admitting of testability. Writers must either know these literary norms before they submit their work or agree to revise their work accordingly; for those outsiders who want desperately to belong, this is a minor concession.

> (Hypothesis: the reinterpretation of popular culture for oppositional ends [Gottdiener 1985] is likely to occur when commercial classification processes characteristic of oligopolistic production systems run up against consolidated status parameters, which generate pressures towards strongly bounded ritual classifications).[46]

Here, the writer propositionalizes his work only parenthetically, almost as an afterthought. He cites Gottdiener as a the guiding authority provoking such an hypothesis. He realizes that nondatabased work needs to assay testable propositions; of course, I doubt that a working empiricist could begin to make sense of his densely neologistic prose in reducing it to discrete variables. Form and figure are important here; the content is only parenthetical. This narrative technique is available to anyone: For example, Sartre might have said, "Hypothesis: If people must die, then life is a useless passion." Science would probably methodologize Sartre more densely: "If individual land-based hominids are physiobiologically prone to fatally entropic states, then we can predict the absence of shared semiocultural value systems."

Other "theory" articles unashamedly display their propositionalization prominently.

> Proposition 1. High-cultural tastes or "cultural capital" assist in developing or reinforcing relations with high-status persons.[47]

This passage is followed by some argumentation both elaborating and supporting it.

> In advanced societies, the arts (high and popular) occupy a privileged position among identity-defining conversational currencies for several reasons, not least of which is their availability. Television provides a stock of common symbols for nearly everyone, and youth-oriented cultural forms pass easily across class and geographic boundaries. The high arts have become important status markers, for they are subject to few barriers of age, region, or gender, and are consecrated in school curricula. Consumption of high culture is associated with status throughout the industrialized world, in socialist societies as well as in the West (Gordon and Klopov 1975; DiMaggio and Useem 1978a). If there is a common cultural currency, the arts (supplemented by fashion, cuisine, and sport) constitute it.[48]

For a science article, this long passage is relatively accessible and interesting. Technical barriers to understanding are few, and the citation of authority is neither oppressive nor ritualistic. The proposition preceding it is banal to the point of absurdity. The more discursive argument of the passage is telescoped into a few lines of text, reducing its original suppleness. There is absolutely no narrative reason why the long paragraph must truncate into Proposition 1, as it is called. Science's literary strategy forces prose into the meat grinder of propositionalization simply to satisfy the methodological requirements of journal science. Propositionalization promises empirical testing, or so it seems. In fact, the propositions interspersed into the slower prose merely serve a talismanic function. The author does not expect people to research them as such; he merely wants to add value figurally to his narrative. The more prose reads formally (as methodologists would say), the less it seems like poetry, politics, perspective.

The next proposition is much the same as the first; it is obvious and unargued to the point of declarativeness.

> Proposition 2. Possession of "cultural capital" is related to positive outcomes in the status-achievement process.[49]

The next proposition is similar except that it is supported with little buttressing argument. The semiotic conversion of propositional prose into textual prose is virtually one-to-one; there is little translation going on, little to warrant the propositionalization of one construction but not the other.

> Proposition 3. Persons with wide-ranging networks develop "tastes" for the widest variety of cultural forms.[50]
>
> Research on social networks has shown that socioeconomic status is positively related to their size, complexity, and diversity, and negatively related to their density and average tie strength (Fischer 1982; Campbell et al. 1986; Marsden 1987).[51]

Not only is Proposition 3 obvious on one level; cosmopolitanism is characterized by cosmopolitan taste. Even the homely word tastes is uttered in quotation marks—"tastes"—to give the impression of technical usage. But surely taste methodologically is equivalent to taste in the ordinary sense. While not a proponent of ordinary-language philosophy (merely ontologizing ordinary unconsciousness as reason, thus vitiating real reason), I am unable to read the parenthetical "tastes" as semiotically different from what one ordinarily means by tastes—preferences in consumption or judgment. This is especially troubling when more technical terms, such as "wide-ranging networks" (what is a network? how wide-ranging?) and "cultural forms," are not presented as technical usages. In the next proposition, the term *genres* is used without differential marking. Yet few sociologists will understand the term genre without technical amplification. And, interestingly, in the argumentation supporting Proposition 4 we find the word *robust* used as a valorizing adjective ("a robust finding of studies"), again raising questions about its intended technical usage.

Finally, journal theory becomes equivalent to the literature review section of journal science when "theory" is reduced to a citation device referring only to other people's work.

"The McCarthy-Zald theory, then, rested on five major propositions."[52] The "McCarthy-Zald theory" is not a theory in the sense that it speculatively reconstructs a whole social universe, as people like Parsons and Marx did. "Theory" here refers simply to a body of empirical work supported by propositions seemingly organizing that research into rigorously formal patterns. "Theory" here is nothing more than what journal science calls the "discussion" section, where the findings are rendered into prose as the programmatic basis for further research confirming, disconfirming or elaborating them. Indeed, anything and everything can be called a theory when theory loses its specificity as globally constitutive intellectual expression. Theory can be as local as the discussion of findings on age of first coitus, or it can be as grand as Weber's presentation of a theory of world-historical modernization.

Journal theory loses its mind when it becomes any prose construction attaching to any body of empirical research. That is exactly what science intends; theory is the absence of figure—table, equation, graph, model, picture. Absent data to be figured, theory is simply words embellishing the central figural gestures of the text. Theory—words—sustains science in between pyrotechnic displays of nonprose; as such, science is the absence of thought. Ironically, disciplinary founders who in a premethodological era used to present the world mediately in concepts and constructs—Aristotle, Locke, Comte—today are called theorists; yet theory is prohibited from speculatively reconstructing world history except in propositionalized banalities. But disciplinary founders were not only theorists; they encompassed a great deal of empirical territory too. Indeed, for them the distinction between theory and science did not exist; methodology was not yet a rule of knowledge. To call these dead thinkers theorists only limits theory to curatorial work. The other kind of theory abroad today is theory construction—where theory does everything but think. The theory-science distinction conceals hierarchy. Today officially accredited academics are prevented from doing the sweepingly constitutive work of the Grand Theorists by disciplinary norms subordinating thought to technique.

Chapter 10 Essaying a Different Text

Essaying Science

I have read journal science as a literary and hence political expression. Facts are not simply facts; science writers, in reporting the world, also constitute it; speculative reason can read science and thus reverse its subordination to it. This much is clear by now: Science as a deceptively ontologizing force only freezing people further into place unravels; methodology solves no intellectual problems.[1] 1 It is rhetoric. As rhetoric, it can be engaged, opposed, even reformulated. In this concluding chapter, I want to essay a different concept of science grounded in that very notion: Good science[2] essays the world unashamedly.

The temptation to rest with ideology-critique has to be resisted; deconstruction's lesson is that it, too, is construction. Undoing science as metaphysical, hence political, narrative is narrative itself. It suggests and thus promotes a different world. The Enlightenment was right to want to enlighten.[3] Whereas myth suffocates, science conceals its own mythic nature in secret allegiance to mythology—a tool of all oppression. Driven by Nietzsche, among others, poststructuralists in opposing modernity oppose liberating modernisms. Reason can reorder history; yet reason is not method but substance. It must be formulated, narrated, argued, revised; as dialogue,[4] reason's process is also its product—Habermas calls this the ideal speech situation, the goal of criticism as criticism. I oppose science in its name; positivist representationality embodied in journal science reflects a world illegitimately drained of history. Science meets theory and thus ensures democracy when it takes account of its own constitutive role, however minuscule in the moment. There is no Archimedean point from which methodology can replace speculative reason. And just because reason itself does not enjoy the epistemological privilege of standing outside history does not mean we should not reason. That there are no metaphysical guarantees in a godless age drives us into the public realm, not away from it. The alternative—abdicating knowledge to a distorted concept of science—only ensures the eclipse of reason.[5] Science never banishes the metaphysical; metaphysics returns in a nearly invulnerable form.

What, then, of written science? Under the sway of discipline, journal science freezes an already chilly social world. The reconstruction of its logic of inquiry as methodology convinces only those who believe that technique solves intellectual problems. Instead, as I have shown, technique is only another poetry, another formulation of possible reason. In essaying science here, I have read science in effect as essay[6]—posturing a world in recommending it. All writing does this. Description suggests a world its reading would enact, whether ineluctably tragic or politically open.[7] Essaying methodology unashamedly suggests a concept of science's narrative as essay itself.[8] We denigrate the essay precisely because we believe science's claims to have dispelled perspectivity, polemic, poetry. This is untrue. Science's version of the world is no less speculative than that of Aristotle, Aquinas, Adorno. My argument is that speculative reason embodied in the essay—literally, the attempt—rests on solid epistemological ground. No amount of authority citation and figural display dispels the inherent ambiguity of the world. Yet, unlike positivist science, the essay does not pretend not to pretend; it argues its metaphysical case as the political theory it is. Narrative recommends.

I have essayed methodology as essay itself. This does not disqualify figural usages—statistics, number, table, equation, graph, model, picture. If anything, it strengthens these expressions by removing their portentous ontological implications. Figural pyrotechnics do not replace rhetoric as a mode of substantive reason. Mathematics does not win arguments; it only argues, embellishes, aestheticizes. This is virtually impossible for science to admit today, for it has enjoyed an exemption from metaphysical examination for too long. Nearly everyone believes that the world is disclosed in number and figure; yet this only numbers and figures the world more, diminishing the role of speculative reason. As essay, science is no less speculative than other discursive formulations of the world. It is no more speculative, either, a favorite prejudice of humanists who oppose quantitative analysis as unpoetic, perhaps even immoral. In this chapter, I want to argue against that neoromantic critique of science, even though I blame science itself for its differentiation from the humanities, from what it constructs as nonscience.

The split between the two cultures concretizes in the rhetorical differences between the scientific report and the essay. One is rigorously representational, driven by methodology; the other is whimsical, perspectival, poetic, sometimes political. Where Montaigne popularized the essay as a legitimate mode of self-presentation for humanists, the scientific research report dominates the journal literature in the natural and social sciences. The article-essay split is false. Research articles are essays, and essays are articles. Articles constitute the world by reflecting it; essays reflect the world by constituting it. The results are the same. In both instances, writers write writing. Words both mediate and constitute. Style, whether methodological or literary, does not solve problems in itself. The burden of proof rests squarely on the shoulders of the author, who must above all portray a credible world imaginatively suggested in proseful and figural displays.

Today in science disciplines the research report dominates academic discourse. Research reports authorize themselves with reference to overarching literatures and establish their representational credibility by figuring their results methodologically. Science as literature differs from fiction in that it must laboriously claim validity for itself; the essay must establish a ground from which it can be believed. Fiction is absolved of this responsibility inasmuch as we never fail to doubt it. Fiction is fictive; the fiction writer imaginatively constructs an unreal world that, she or he hopes, so resembles the one we experience that in the course of the story we forget our difference from it. The fictive world gains verisimilitude simply by doing good work.

Science methodologizes its account in order to establish the credibility of its report. Without its first page, literature review, technical excurses, and then "discussion" at the end, science would remain as speculative as fiction. Its world would be no more trustworthy than that of Camelot and castles. Yet we forget that science and fiction lie along an unbroken continuum. Both reporting what they intend to be a truthful world, they use different voices and strategies for capturing the reader's trust. Whereas fiction builds a believable world inhabited by common-sense experiences we can recognize as similar to our own, science gains instant credibility by presenting itself as a methodological artifact, its findings gained through appropriate technical strategies of inquiry. But methodology is only the literary reconstruction of inquiry, not a preontological arbiter of all disputes. In this sense, methodology is science's way of accustoming the reader to a truthful report. The disciplined reader recognizes a disciplinary account as a legitimate one; assuming that the opening literature and methodology sections have been presented competently, the reader has faith in the evidence presented. Bad fiction and bad science both fail to conjure a believable, recognizable world. They do not successfully entice the reader into the fictive world of literary artifice.[9]

Essayists as humanists are quite comfortable with the nature of their craft. The essay stands midway between fiction and science as we divide them today. I argue that, in effect, the three are the same, inasmuch as they all construct a believable (or unbelievable) world without methodological or metaphysical guarantees. Essayists recognize their affiliation to the literary project; as such, they compose a world without pausing to claim methodological sanction. The essayist is unashamed of his or her authorial presence in the text. After all, what makes an essay essay a world is precisely the circumstantial ground in authorial experience making way for a believable account. The essayist recognizes the utter contingency of writing (about) a world eluding any reduction of it to concepts and images. Realizing that the mediateness of words is as much a literary resource as an impediment to clairvoyant understanding, the essayist, like the novelist, does not worry about being used by the words he or she uses. The essayist realizes that there is no point outside the essay from which a truer vision can be gained. The essay acknowledges its own ellipses, omissions, even distortions and thus provokes other accounts (including its own) as re-visions.

As such, then, the essay is literature[10] that refuses to suppress the author's perspectivity and passion. The essayist recognizes that science is an essay too—simply that, an attempt to disclose a fundamental world. As often as the essay succeeds in provoking thought, it also fails. Either its language is too private, its prose too leaden, or its imagination too limited to spark insight and thus provoke other literary versions of it. The essay is not only comfortable with the corrigible nature of its account; it celebrates this corrigibility as the possibility of other accounts. Without them, the essayist remains enclosed in a private authorial subjectivity, having touched and moved no one. Good essays—good science, here—prompt responses, arguments, different versions as the literary program of a nonviolent, dialogical community. From Plato through Marx to Habermas, philosophers have located the good society in conversation. I hold with that; here I want to suggest a conversational approach to science considered merely an attempt—essay—to enliven a larger literary community.

An account essaying science does not require that we dispense with the hard-core technical apparatuses bulking the science paper today. After all, it is a humanist prejudice that essays are first-person ventures into leisurely reason. While they can be that, too, essays are not exhausted by their folkways today. Just because we usually take essays to be halfway between straight fiction and high science—neither fish nor fowl, chatty but unrigorous—this does not require us to disqualify heterodox essays out of hand. If science is already a literary attempt, as I think it is, it is not our right to force it into a peculiarly narrow version of writing to which the narrative norms of methodology are unavailable. If we deny science methodology, we accept science's claim that methodology is, after all, different—not a speculative text but an Archimedean apparatus for resolving philosophical and empirical muddles. Methodology is no less epigrammatic than the most lilting prose or affecting poetry. It tells no story other than itself; its access to the universe is no less mediated than the access granted to words and image.

This is the real problem with scientists' literary self-conception (or lack of it). They just do not understand or admit that their pyrotechnical displays are merely rhetorical devices, nothing more—or less. Humanists, for their part, shrink at number as an illegitimate truncation of reason (Habermas's own position in *Knowledge and Human Interests).* Yet, ironically, those same humanists pepper their writing with technique and figure, defending shape as the nondiscursive form of reason. We can overcome this mutual incomprehension by reading number as figure itself and then by essaying figure into prose formulations. This does not mean that we must tediously translate every figure into so many words (or vice versa) but only that we allow a multiplicity of formulations—including formulations—as vehicles of reason. There is nothing sacred about methodology; nor is it demonic. Instead, method is only the reconstruction of one's logic of inquiry; as such, it is a vehicle of authorial self-consciousness.

For humanists to recognize the poetic grounds of methodology would enable them to read it and use it. As oppressive as number frequently is in a numbered, patterned world, there are times when we must count and figure (and not only at the supermarket or automobile showroom). Number read rhetorically discloses a world it would freeze in its serializing account. A classical example of deconstructive reading here is Marx's essay of money in *Capital*. As a text, money both conceals and reveals a whole social world; it also helps reproduce that world to the extent to which it is not read as the essay it is. It is simply self-indulgent to refuse the "language of the oppressor" when, in fact, that language oppresses the world over. And one must reformulate the oppressor's discourse both to challenge it and then move beyond it. Reading is political practice to the extent to which communication today fails to communicate—even blocks communication.[11] To dismiss this discussion as mere ideology-critique is another self-indulgence. Ideology materializes domination.

Yet the humanist aversion to methodology is not my main concern here; humanists are few and far between. For every undergraduate student who writes a senior thesis on Chaucer, there are hundreds who major in business, economics, psychology, political science—some of the central disciplines of oppression today.[12] Scientists control the journals and hence the disciplines. They believe that methodology, in itself, solves intellectual problems, thus further reducing the scope of speculative reason. Their journal science only freezes an already chilly world by pretending to stand outside of it in a relationship of representation. Science suppresses its rhetorical nature as ontological and political recommendation precisely to bring about the world disclosed in its forbiddingly technical reflections. Methodology refracts and thus helps to reproduce a patterned, figured world. While not the most crucial vehicle of oppression today, positivist journal science both blunts reason and, in proliferating itself, lavishly fails to do more useful work. Positivist representation, in concealing its constitutive nature, constitutes all the more effectively.

Science values methodology's alleged distortionlessness. The abundant figure in science papers promises a world transparently reflected into the reader's eye (and thus body). But methodology constitutes the very world it pretends only to reflect. The numbered, figured world begins to look like its representation. The posit of social laws only ensures them. But science does not scientifically address its own constitutional contribution to history. Everything is essay—thus disqualified—except methodology. Science disciplines have use for poetry and essay only in their places—across the campus in humanities departments, galleries, museums, the theater. Figured and presented in dense technical language, methodology alone stands above the fray, cleansing data of the passions of perspective and politics. In this sense, I want to approach positivist value freedom as a literary strategy, not simply as a set of methodological protocol statements or deep assumptions. Objectivity is objectified on the science page in signs pretending only to disclose, not also constitute, a world. But representation cannot avoid mediateness; both words and figure reconstruct (and thus reproduce) their objects. Read correctly, methodology passionately recommends a certain order of being—a numbered, figured world in which the scope of freedom has radically diminished. Methodology conserves by mediating, thus constituting, the given world. We must not take methodology at its word; every textual gesture ontologizes, politicizes, supports, and subverts.

Of course, as long as methodology resists an account of its literary nature, it gains value as a representational account. It is impossible to overestimate the self-reproducing nature of discipline; virtually everyone in academic science departments accepts representationality as the goal of science writing. And outside the sciences, most accept science's methodologism as a textless practice. The two cultures differentiate because few recognize and resist that differentiation as the hierarchy it is. Science is bound to hold sway in a scientized world; poetry is added only as a moral and spiritual leavening. The humanities inevitably become the second culture when science postures methodology as a nontext. Science convinces humanists of its difference by appearing to cede meaning-constitution to the expressive disciplines. But this is illusory; science is the real poetry of the age. Poetry is merely an expressive embellishment, lacking in constitutional efficacy precisely what it unknowingly concedes to science.

Reading science as a literary text, even as ontological argument, has a number of advantages. We end the spell of methodological representation, insisting that method is only another rhetorical attempt. Positivism is the central ideology of advanced capitalism and state socialism. We thaw the world science chills; as such, we can learn from science ironically. We also empower different sciences as legitimate versions in their own right. Essaying science opens the possibility of different accounts of the objective world; positivism only appears to exhaust systematic knowledge. I do not want to save science from its worst nature but to interrogate the possibilities of science as a narrative strategy. The problem is that we do not know what science is (or can be) today; positivism monopolizes the available dialogue opportunities.

This is a very serious problem from the point of view of deconstruction and thus political imagination. It is nearly impossible to think a different concept of science outside of its current organizational and literary conventions; "science" inevitably conjures up images of thoughtless number-crunching lifeless prose, metaphysical and thus political conservatism. When one proposes a new science, humanists reach for their guns, or their quills. This is symptomatic of the hegemony of positivism; there is nothing about science compelling thoughtlessness. Deconstruction typically founders because, by its nature, it lacks utopian imagination; working so close to its object, it tends to accept its object's concept for itself. Thus science becomes the "science" conventionalized in the journals, classrooms, conferences. The negation of science suggests an entirely different practice.

This, then, is the topic of my concluding chapter. Having essayed positivist journal science, I want to essay a different science without indulging in programmatism: "We should..." Part of my problem is inherently literary (some might call it dialectical): We must already write in the voice we recommend. Prescription and description merge when we think of them as converging literary voices. I cannot prescribe new science without having come squarely to grips with the present literary practices of what passes for science, especially methodology. What I propose (and exemplify) in the way of new science will be heard as counterpoint to the positivism I have tried to essay; all of this might suggest a different text: Negating the bad overcomes it while preserving it.

Where will this new science come from? Having essayed methodology as a rhetorical strategy of its own, I would suggest that type of reading already as a new form of writing. Some might call it deconstructive, others dialectical. In bad times, constitution will be counterpoint; nonetheless, counterpoint constructs, if only in the absence it suggests, a better world. The critique of deception clarifies by revealing. There is no other way; one cannot just do better science at a time when to do science compels conformity to the literary mode of journal science. One must engage with bad versions of writing before—and, dialectically, as a way of—doing better versions. Ultimately, the difference between critique and constitution fades. Criticism constructs and construction criticizes.

In the meantime, one can only be prefigurative, hoping that one's critique models a better form of writing and thus of community. That the critique of representational science is not the whole story does not mean that we can avoid it, especially when discipline reigns nearly unchallenged in the age of academe.[13] In disclosing the metaphysical and political deceptions of positivism, one necessarily illuminates a more fluid world as well as a language strategy required to achieve it. Avoiding the fetish of the textual, one can still address textuality as a crucial component of domination especially at a time when science uniquely presents itself as a privileged discourse unsullied by the perspectivity, poetry and passion of political writing. Science dominates[14] by arrogating to itself all discourse opportunities pertaining to objective social analysis. In the process, science methodically conceals its own constitutional contribution to the world whose representation it suggests on the flat journal page. The problem immediately before us, then, is the way science disqualifies other versions—the essay, writ large. Science, to be science, necessarily constructs a realm of nonscience to which it claims silent superiority. It is from that realm that the effort to essay science springs politically.

Nonscience: The Return of the Repressed

Science constructs the notion of two cultures, humanities and hard science, precisely to differentiate itself from the inferior culture. No matter how many Chaucer scholars a university can boast, there are always more academics engaged in the technical studies, from law and engineering to sociology and economics. The notion of two cultures stratifies by differentiating; we are to take humanities to be the lesser world.[15] As such, then, science implies the inferiority of different literary approaches to the world; in suppressing its own essay character, science disqualifies essay on epistemological grounds. Methodology implicitly differentiates itself from the nonmethodogical, not only the bad methodological.

For journal science, the nonmethodological is called different things; the meaning is always the same. Mush, qualitative work, armchair speculation, thought pieces, essay, humanism—all these terms, and more, characterize science's constitution of a realm of nonscience beyond it (thus beneath it). Without a second culture against which to orient itself, science would jeopardize its own imperial claims. Flattened onto a long epistemological continuum, as I try to do here, scientific methodology loses its Archimedean character in becoming merely rhetoric. Of course, rhetoric does not have to apologize for itself when it makes its case well. One can do good methodology—methodology that recommends and convinces. Yet science inferiorizes the essay precisely to demarcate itself from the problematic speculativeness threatening it with undisciplined, unprincipled—wild—knowledge. Science claims to be on high ground, whereas humanists swim in the vast sargasso sea of authorial subjectivity.

While scientists would admit that "there is a place for" the humanities, that place is off the main academic highway. The move toward a "new" liberal arts sweeping the United States in the wake of conservative laments for the (sexist and racist) good old days reflects the marginal status of the humanities. Although not completely banished from the academy, humanities disciplines such as English are remedialized; illiterate high school seniors are thus rehabilitated by the time they graduate from college with degrees in the hard sciences and professions. In my university, English literature is taught "for" management majors, not for its own sake. Similarly, my own discipline, sociology, is offered to just about everyone but sociology majors (of whom there are few).

The notion of two cultures implies separate but equal intellectual domains. Instead, essay and performance disciplines languish outside the scientific-technical mainstream. Not only do they have low status; they are remedialized and thus further reduced. It is impossible to understand the rhetorical relationship between science disciplines and essay disciplines without also understanding their organizational relationship in the modern university. Science disqualifies the essay as nonscience and thus only compounds the marginalization of essay disciplines in the socio-organizational setting of the university. Science becomes science by playing itself off against the speculatively weak-kneed essay. Methodology differentiates science from texts and departments in which authorial subjectivity is not only revealed but celebrated. Methodology makes the difference between two cultures an epistemic as well as organizational issue.

Ironically (again, some might say dialectically) the science-essay contrast undoes itself at the margins. Science's construction of essay disciplines as nonscientific makes way for science's own reformulation as essay when humanities disciplines, under siege by science, have already methodologized themselves. The falling status of the humanities in the late-twentieth-century American university, along with the deepening critique of the slavishly scientific-technical nature of contemporary intellectual life, has issued in the rise of literary and social theories capable of comprehending the denigration of their marginality as symptomatic. Critical literary and social theories of the text have sprung from a burst of European scholarship inspired in different ways by Marxism, feminism, psychoanalysis, and semiotics. Intellectual figures such as Adorno, Habermas, Derrida, and Foucault suggest an approach to textuality rooted both in literary and political analysis. A massive and heterogeneous attack on positivism, these theoretical developments have spread like wildfire through a variety of humanities and even a few social science disciplines; indeed, this study is scarcely imaginable without reference to some of these figures.

Literary as social theory methodologizes textual analysis as cultural and political criticism. Against the positivist New Criticism taking each text on its "own" terms (whatever that means), these deconstructive approaches to criticism interrogate the relationship between text and world, construction and criticism, writer and reader, and text and subtext. Science's own self-reconstruction as methodology is now matched in humanities disciplines by the reconstructive and deconstructive methodologies capable of attacking the epistemological and hence political assumptions of a positivist culture. While different people phrase the politics of literary theory differently (some more or less Marxist than others, for example), literary theory essays cultural discourses—such as, here, science—as constitutive, not only representational. Culture itself is reformulated as a thoroughly material practice, not as epiphenomenal (Marx) or merely imitative (Vienna Circle).

The methodologization of cultural criticism dialectically issues in a methodical critique of positivist methodology as a literary text in its own right. Nonscience fights back by refusing to accept science's stratifying differentiation of itself from the speculative studies. Today, through the Frankfurt School, deconstruction, and feminism, the text is read as a world (and world-constitutive); while the text does not exhaust reality, neither is it outside or above history. Science, in freezing history, contributes to it. Grounded methodologically in a variety of assumptions about the nature of cultural signification, deconstructive reading essays different discourses such as science. In this way, intellectual approaches disqualified by science as nonscience (if retained institutionally in remedialized forms) address cultural discourses such as science from the outside. Deconstruction refuses to concede the possibility of purely representational knowledge. Instead, scientific discourse is read to constitute an object domain and thus a world in its own image. Scientificity is a construction of science itself, protecting science's epistemological and institutional status.

From the outside, then, science reads like an essay. Its methodology, reduced to gestural devices heavily reliant on number, is read as a persuasive text, nothing more (or less). Science is drawn into the heterogeneous discursive practices unable to authorize themselves from the outside. Every methodology vitiates itself when it seeks absolute grounds outside of rhetoric. Deconstruction and critical theory best positivism precisely because they make their own centerlessness thematic; nothing guarantees knowledge a priori. Credibility is the only standard of deconstructive knowledge. Essayed knowledge takes its chances in the heterogeneous discursive worlds of the time. Essay itself becomes political practice when it builds community in the talkative nonidentity of many versions; it might be said that the best writing provokes the most gentle debate and dispute. The worst writing remains monologue, thus disempowering.

Resisting the imperialism of positivist representationality, my approach to cultural criticism overturns the two cultures distinction as a positivist artifact. Either there is exactly one culture—discourse—or there are many—discourses. The natural and cultural studies do not differentiate around the issue of methodology; even literary theory can recontruct its readings on methodical grounds (e.g., deconstruction, critical theory). In disqualifying certain discourses as nonscience, science invites its own interrogation as a power center. In failing to essay its own scientificity, journal science convinces only itself. Everyone on the outside reads science as the driven allusion it is; science suppresses the mediateness of the world precisely to bring about a world order in which mediateness—freedom—is (freely) abandoned.

Methodology has become a habit. Journal scientists script it into their papers on the assumption that readers and editors require it. Methodology-less papers are dismissed as essays; recently, a book publisher rejected a manuscript I submitted as a "thought piece." I had hoped it was thoughtful. Outsiders puzzle over the figural displays rapidly displacing words in journal science. We suspect that the markings signify something; thus, as here, we attend to them as text. Whereas scientists reject the essay as nonscience, many humanists are tempted simply to dismiss science as a nontext. Both are wrong. Science is an essay, and essays can be science. The issue is not form but content. One can learn about the world from poets, philosophers, statisticians. Methodology suggests a patterned, figured, numbered world; understanding it that way only entrenches it if methodology becomes ontology. Instead, I essay methodology as essay—understanding but also opposing a certain order of things.

Humanists often suggest that to write science inherently alienates reason by abstracting from experience; instead, one must give experience its due. But number and figure can be traced to the pretheoretical experiences of people in a commodified, monied, edificed world. None of us is a stranger to the rule of number and figure over reason. To understand the rule as a historical accomplishment of capitalism and state-socialism helps undo it, instead suggesting dialogical rules of corrigible expression as alternatives. Counting is immoral only in a society in which no one is counted, aggregated, caused, fragmented. Today people submit once to the rule of abstract equivalence; they are their unwaged and underwaged labor power. They submit twice when they read their fates from representational texts chilling experience into fate. The mathematization of the world is essayed by texts pretending only to copy nature, not constitute it. Only essay can undo essay; it is impossible to avoid using its language, for that is the discourse structuring people's lives.

Humanists can either reject science as an alienation of spirit or they, too, can claim science. There is no middle position—two cultures. Neo-Kantian differentiation only stratifies; the deconstruction of duality reveals superordinate and subordinate terms. The rejection of science takes the conventions of positivism to be the essence of science—computers, statistics, survey research, literature reviews. While humanists understandably would avoid a science denying them epistemic validity in their own right, the only way out is through; positivism must be shown to be a literary production and then produced differently. The stakes are objective knowledge. As ontology, objectivity freezes over our fate, thus secretly reproducing it. While objectivity exists in history, susceptible to reformulation, it does exist; the poem does not end the nuclear threat or stop starvation. Yet poetic understandings of the discourse of objectivity allow that very discourse to provoke a different world in its example of itself as a corrigibly different text. In understanding objectivity as text, we empower readers to become writers.

Science's differentiation from subjective knowledge only renews science's disciplinary power. It is no accident that methodology is taught as professional expertise, not as a universal cognitive capacity. In privileging methodology over speculation, we privilege the people who practice it. Scientific-technical expertise backs political-economic concentration; it authorizes elites and disempowers the uninitiated. The deconstruction of methodology merely as rhetorical style democratizes it; the persuasive skills associated with it are no longer viewed as mysterious.[16] A better science undercuts disciplinary and thus corporate and state power when it is essayed on behalf of universalistic interests. In exempting itself from history, science privileges a certain kind of history. Once science loses its exemption from history, it no longer possesses the capacity to ontologize—inequality, oppression, technocracy. Instead, ontology becomes a generic intellectual competence; indeed, as Plato knew, the constitution of being constitutes good being itself.

The aversion to methodology is a general one; difficulty dissuades us from mastering it. Thus we tacitly legitimate the rule of supposed experts enjoying access to recondite truths and the means with which to express them. But anyone can figure and number writing. Methodology suggests its own mysteriousness precisely to keep outsiders out. Of course, everyday science supports the impression of its own esoteric nature. The casual reader can scarcely make sense of the technical esoterica and figure without advanced training. But the definitive antidote to positivism is not statistics courses for socialists and feminists (although that is not an illegitimate aim in its own right). Instead, we must learn to read methodology as the rhetorical text it is; while a certain baseline knowledge is required to make sense of multivariate analysis, a grounding introduction to social statistics and research methodology can be gained quickly and then refreshed in the course of deconstructive reading.

Methodology mystifies way out of proportion to its intrinsic difficulty. Its code is easy to crack once we treat its constructions as literary conventions often arrayed together haphazardly. Humanists shrink from the implied difficulty of the science page. This is precisely what science intends. But outsiders can essay science by understanding it as narrative—a pastiche of gestures and signs connoting and thus recommending a certain order of being. Methodology loses its ontological hold precisely to the extent to which we deprivilege its implied discursive claims. Once we realize that methodology resolves no arguments outside of the contingencies of conversational ambiguity and irony, then we are free to learn from the figural and technical work of science. That is the important thing: Positivist science discloses much of value about the empirical world. It is irresponsible to ignore their discourse, even though they disqualify essays such as this out of hand. One must criticize positivism as a literary and hence political strategy; one cannot ignore it.

In examining methodology from the outside, we resist the temptation to shun objectivity altogether. We also refuse to allow science to share the intellectual world without acknowledging its literary nature. Positivism bursts the shackles of its bondage, relentlessly disqualifying nonpositivist studies as nonscience, thus nonsense. Positivism refuses to coexist; the humanities share the field with it only at their peril. Positivism constructs the humanities as nonscience out of its own epistemological imperialism. In so claiming science for itself, positivism ensures that the humanities will reject science as such. For humanists, the only utopian alternative is the nonscience enforced on them by a positivism illegitimately confusing itself with all objective knowledge.

In claiming all objectivity, positivism leads humanists to relinquish objectivity for subjectivity. This is not a fair exchange, especially when subjectivity is already objective. Absolutism thus breeds another absolutism. The science-nonscience duality, constructed by an imperial positivism, repeats the standard either/or of bourgeois society. For critics of positivism, this is a troubling dilemma. The conflation of positivism and objectivity risks losing objectivity altogether. Although evocative, poetry is not the universe of possible knowledge either. While poetry cannot be disqualified prima facie, neither is its allusiveness a sufficient address to objectivity. In typifying nonscience by poetry, science ensures that critics of science retreat to expressive subjectivity when confronted with an imperial positivism.

Thus positivism in constructing alternatives to itself as nonscience ensures its own hegemony. The university has a place for Chaucer scholars if not for dialectical materialists and socialist feminists. Criticism betrays itself by succumbing to its own marginality; while its marginality is a fact, it ought not to be accepted quiescently. Struggle is difficult and frequently painful; capitalism institutionalizes conflict and thus leads reasonable people to avoid it at any cost. But the epistemological claims laid by positivism must be resisted. Science both narrates and constructs a world it pretends to reflect presuppositionlessly. A new science is born of its critique of the old; as here, critique resists its own construction purely as critique. Critique also constructs, exemplifying a more general liberation. Here critique must neither distance itself from literature nor renounce objective knowledge; science in grasping the objective also changes it.

Writing Difference: Beyond Authority and Methodology

Too much criticism labors under the weight of the old and the present. Either we grasp at slender utopian threads from Marx himself or we descend to piecemeal agenda setting. But the answers are not there. Marx did not speculate a new world sufficiently to suggest new epistemological practices (although there are powerful hints—e.g., the 11th "Thesis on Feuerbach"). Nor is agenda setting worth the paper on which it is written. It is a virtually irresistible temptation for critics to propose minor modifications of the present unjust social arrangements, installing themselves in the chairs of power. But the answer to our problem is not simply to change editors at the mainstream academic journals, start new journals, even reform the university. All these things would certainly make life more bearable—more intelligent—for marginal intellectuals. At least academic jobs and tenure would be more attainable. Yet journal science simply reflects a deeper positivist world in which representation surreptiously constitutes and thus reproduces.

Nor is the avoidance of representation an answer. Humanism founders on its good intentions. As sheer evocation, cultural expression fails to comprehend the complex totality enmeshing it in multiple layers of significance. Poems matter, and yet they fail to theorize sufficiently. Of course, this implied notion of theory is itself scientific; I model knowledge as the approach to objectivity. Positivism is so pernicious because it pretends to exhaust systematic knowing. That it fails—that we can read science in a literary and political way—is cause for optimism and then for reformulation. We can narrate science differently and thus in our small way change the world to which science is a constitutional address. Showing people deconstructively that the patterns figured by positivist science are false—that history opens from the text outward—is sound political practice. While textual critique is insufficient, it is no more insufficient than labor organizing or rape crisis centers. Everything matters to a world so complexly interwoven that it takes enormous conceptual work to comprehend the modal power principles as well as to locate interstitial pockets of possible resistance. Foucault shows that power is everywhere; no discourse is absolutely constitutionless, even—especially—science.

The irony of different writing is that in wresting away its identity from the dominant discourse, it trades too much on convention. The left citation of authority and left methodologism are no improvement. Yet critique cannot ignore its object—here positivism. New science only emerges dialectically in tension with the old. Critique prefigures a better discourse and thus a better world in its own example. Yet its example necessarily copies from what it opposes; hence left critique is frequently as authoritarian and methodologically compulsive as mainstream authoritarianism and methodologism. Sometimes left thinkers defend this on grounds of legitimacy. Yet books with footnotes will convince no one not already convinced; footnotes change no minds. The problem of rhetoric is always and only strategic: What it takes to change the world is what it takes; there is no sense imitating the old if imitation only reproduces it.

Different writing out of what it takes to be strategic expediency repeats positivism in opposing it. Yet in marking its difference from journal science, different writing risks losing public discourse altogether. An example of the first is orthodox Marxism, where fealty to the original canon results in a debilitating scholasticism and academicism.[17] An example of the second is the methodologization of deconstruction in English departments.[18] Today, academic Marxism stifles intellectual innovation under the heavy weight of disciplinary discourse. And Derridean wordplay fails to grasp the social object. This does not mean that we can simply abandon Marxism or deconstruction for the insights they afford. Precisely the opposite; for me, they are integral to the emancipatory project. Yet different writing falls between two stools: It must disentangle itself sufficiently from disciplined discourse to be able to breathe freely and yet it must avoid epistemological anarchism dressed up in Nietzschean (or other) guise.

Liberation—here, new texts, new science—cannot be formulaic. Programmatism self-contradictorily writes in the voice it recommends, as such prefiguring new discourse successfully having vanquished the bad discourses of the present. What this takes epistemologically and theoretically cannot be known a priori. Instead, the objects of our domination must become our subjects—here science, later money, edifice, and figure. Imagination imagines a better world in some ways continuous with the present one. The notion of dialectical leaps is as ungrounded, as metaphysical, as the liberal faith in linear evolution. Marxism is millennarian in spite of itself. Nietzsche reminds us that history is meaningless; but that does not mean we cannot give it meaning. History is a context within which freedom unfolds as determinate negation and construction—frequently the two together.

In itself, difference only differentiates, thus failing to liberate. As Derrida shows, differentiation frequently conceals stratification, the polar terms of duality falsely posed as equivalents (e.g., subject-object) and thus really constituting homogeneity (where the stronger term constitutes and thus distorts its negation). Derrida thus poses difference (or, as he calls it, *differance)* as the fulfillment of what duality only falsely promises. Yet difference in itself does not liberate if it fails to address the differential power imbedded in the discourses of dualism. The fact remains that the stronger term—here, Western positivism—controls the discourse so fully that it can name the marginal term (e.g., nonscience) and thus disqualify real marginalia out of hand. Positivist difference is false difference; thus difference must insist on its right to name itself different, enabling it to subvert dominant discourse from the outside.

A formulaically different science risks being too much like the science from which it claims difference. We can and should enunciate principles of difference from now until doomsday—more authorial involvement in the text, more dialogue, less unexplicated figure, less reliance on leaden authority. All these things are important in themselves. Yet different writing solves nothing if it does not also suggest (and thus prefigure) a whole different social world within which textual significance has significance for a wide variety of material practices beyond itself. A different science must be sketched and then practiced within a notion of a different totality. Yet the totality cannot be envisaged or achieved beyond the particular practices constituting it—a revolutionary hermeneutic circle. The transformation of journal science can suggest (and in turn be suggested by) transformations of all sorts of human and human-nature relationships. All of this must happen simultaneously for the totality to change.

Difference must not only be different—here, nonpositivist cognition. It must also challenge its subordination by a dominant term alleging to embrace difference within itself. After all, mainstream academic journal editors claim to embrace a host of theoretical and methodological perspectives. This is untrue; the differences are concocted out of sameness in order both to show pluralism and to preempt real difference. Taking both epistemological and organizational forms—journals and academic disciplines—positivism will not concede legitimacy without a fight. Different writing must do more than lay claim to heterodoxy. It must also challenge the positivist discourses themselves as political practices surreptiously reconstituting a positivist world. Today construction proceeds through critique. Derridean difference deludes itself that its mere existence apart makes political headway. While survival is obviously better than capitulation, autonomy is less political than resignatory. While distance preserves mind and body, more must not be claimed for distance in the way of political transformation. Whether or not autonomy constructs is always an empirical question; simply to assert difference is unconvincing in the abstract.

Under the sway of instrumental rationality, the pragmatic left will ask how—exactly—this discussion issues in different discursive and organizational practices. I can only point to this book as a prefigurative example of what it suggests programmatically. That is not a lot. Yet my attempt to essay science suggests science itself as a kind of essay, unashamed of its literary gestures simply as rhetorical strategies. Science is a shibboleth only for those who accept its claim to exhaust knowledge. But poetry is science, too. And science poetry. My literary reading of methodology preserves the possibility of empirical analysis that proceeds on the basis of speculative reason, not technique. This book is both against science and science itself. As an essay, it suggests essay as a rhetorical form for empirical and theoretical analysis. Of course, to say this already risks the obsolescence of the term "essay" as a slogan; if everything is essay, nothing is. Yet today the term "essay" trades on a notion of essay as the attempt to persuade. Reading science as rhetoric in this sense suggests a rhetorical concept of science and thus opens science to all sorts of writers and versions.

In this I return to the notion that in interrogating the good, science itself can constitute it—Plato's republic to Habermas's ideal speech situation.[19] In its perspectivity and sense of metaphysical openness, good science constitutes the good by provoking a dialogic community. If we understand empirical analysis as political argument for the world implied in it ontologically, then we purposely make way for a whole host of versions none of which is silenced by either authority or methodology. Epistemological democracy opposes authority and methodology as obfuscations of every scientific text's constitutionality. Even—especially—journal science recommends a whole order of being in the figural gestures littering its pages with nondiscursive markings. As such, the rhetorical nature of journal science itself deconstructively contradicts its avowed reliance on authority and methodology as pillars of its reason. Positivist science narrates and hence constitutes a world ruled by authority and methodology; but as essay itself, it betrays its own claimed representationality.

At least positivism poses a standard of scientificity—sheer representation. I risk standardlessness when I collapse the categories of literature and science into essay. Yet the standard of knowledge is objectivity; whether or not the standard is met is an empirical question, answerable only in the lines and gestures of the text. *Moby Dick* and the oeuvre of Talcott Parsons have exactly the same epistemological status on this view. Methodological conventionalists would respond that Parsons writes about the "real" world, Melville about an imagined one. Yet humanists and critics of Parsonianism would reverse this, claiming more verisimilitude for the novel than for Parsons's oeuvre. Procedural criteria—citation of authority, methodology, academic employment—do not decide truth claims any more than seeming fictiveness denies literature. That Moby Dick (the whale) may never have existed is beside the point if Melville conjures a believable world in which whales and people behave the way he says they do. After all, which sociologists have ever seen a unit actor?

Methodologists cry anarchy when their procedural principles of truth are deprivileged and decentered. I respond: Fight essay with essay; if an account is weak—bad science, bad art, bad ontology—reformulate it with reference to the wealth of experience available to writers. Scientists must realize that they are writers too; their worlds are no less imaginary than the fictive worlds of novelists and poets. In essaying every text, essay becomes a mode of rhetoric adjudicable on any number of grounds—aesthetic, political, personal. Essaying discourse does not preclude evaluation but rejects procedural criteria of evaluation; neither authority nor methodology settles arguments except rhetorically, as persuasive constructions of the imagined world.

Far from vitiating evaluation, the essay of discourse makes way for many voices and versions, not only the vouchsafed accounts of credentialed scientists or well-known artists. Authority and methodology disqualify too many competing accounts, giving the lie to the liberal notion of the marketplace of ideas. Both the epistemological and organizational contexts of science today create enormous resistance to innovative ideas; one lacks methodology, a literature review, sponsors, academic employment, tenure, and the like. But intellectuality is an emergent property of thought, not a procedural attribute. The more intellectuality is a brand bestowed by institutional and intellectual affiliation—the more it is conventionalized—the less people are encouraged to be intellectuals, thus to participate effectively in the public sphere.

In collapsing the categories of fiction and science into essay—speculative reason—I shatter the false idols of authority and methodology. I may also risk too much dedifferentiation; surely many will lament the absence of the standard distinction between art and science. Although distinctions are not bad in themselves—indeed, they can preserve the autonomy of the inferior term—the differentiation of art from science suits science's needs as the superior term today. If art and science are only separated gesturally—perhaps the extent of figural expression—art is suddenly raised to a higher epistemological and thus organizational level. For its part, science is democratized when the rules of authority and methodology dissolve into the persuasiveness of rhetoric. When every text is essayed, the true arbiter of validity claims becomes reason, not methodology or authority.

In turn, this has enormous consequences for the way we think about and organize the academic discourses revealingly called disciplines. When art and science enjoy the same epistemological status, we open their different institutional embodiments to critique. Some argue for the dedifferentiation of knowledge as a way of resolving intellectual and institutional fragmentation. I would argue for the interrogation of differentiation on literary and political grounds; difference debilitates when it conceals hierarchy—the sciences *over* the humanities (and thus over nonpositivist versions of themselves). I end with power. What makes science science today (and what devalues and disqualifies other accounts of the world) is a certain epistemological and institutional privilege. The Enlightenment supplanted religion with science but thereby deprivileged speculative philosophy, the traditional bulwark against mythology. The problem is that the original philosophical problems—justice, truth, beauty—bedevil the modern world as never before. People have even invented a name for a philosophyless age—the postmodern.[20]

Ironically, science did not replace religion, which thrives in these dismal times. And in undercutting speculative philosophy as methodologically rudimentary, science lost the only basis on which it could resist its own tendency to become myth—scientism. Intellectually, at least, we now have the worst of all possible worlds: We disqualify reason as archaic; religion seems to redeem postmodern anxiety; science drains into proceduralism—the rituals of authority and methodology. In this context, it is tempting to prescribe single nostrums—more of this, less of that. The call for "essay" reads like such a nostrum. We must avoid that, for words tend to reify into slogans betraying their provocative vitality; I essay science simply because the two words form an enlightening, perhaps disturbing, oxymoron. We know that solutions are not single or simple; but some of us suspect that we are rapidly approaching a time when it is illegitimate even to talk of solutions. In countering religion, science loses its own mind. If not the only problem confronting Western civilization, that problem alone makes it exceedingly difficult to think practically about what to do next.

Notes

1: Reading Writing Differently

1. This suffocation of thought has recently become topical with the growing concern about the university's role in culture and politics. For opposing views, see Allan Bloom, *The Closing of the American Mind* (New York: Simon and Schuster, 1987); and Russell Jacoby, *The Last Intellectuals: American Culture in the Age of Academe* (New York: Basic Books, 1987). I have entered this debate on a more theoretical level in my *Fast Capitalism: A Critical Theory of Significance* (Champaign: University of Illinois Press, 1989).

2. Marx's original critique of ideology was elaborated in his and Engels's *The German Ideology* (New York: International Publishers, 1949). Since then, a number of people have reformulated his theory of ideology. See Georg Lukacs, *History and Class Consciousness* (London: Merlin, 1971). Also see Herbert Marcuse, *One-Dimensional Man* (Boston: Beacon Press, 1964); and Max Horkheimer, *The Eclipse of Reason* (New York: Seabury, 1974).

3. Building on critical theory and deconstruction, among other traditions, I have elaborated a version of ideology-critical deconstructive reading, or what I call disciplinary reading, in my *Socio(onto)logy: A Disciplinary Reading* (Champaign: University of Illinois Press, 1989). Innovations in theories of interpretation since World War II have contributed much to the original Marxian critique of ideology. See, for example, Jonathan Culler, *On Deconstruction* (Ithaca: Cornell University Press, 1982); Michael Ryan, *Marxism and Deconstruction* (Baltimore: Johns Hopkins University Press, 1982); and Terry Eagleton, *Literary Theory: An Introduction* (Minneapolis: University of Minnesota Press, 1983). These works and the traditions they describe and deepen add significantly to standard work in the sociology of science. Indeed, this study eschews traditional approaches to the sociology of science in its standard institutionalist garb as well as its discourse-analytic reformulation by quasi ethnomethodologists. See, for example, Robert K. Merton, *The Sociology of Science* (Chicago: University of Chicago Press, 1973). Also see his "Priorities in Scientific Discovery: A Chapter in the Sociology of Science," *American Sociological Review* 22 (December 1957): 655–59. Marxists have also elaborated a version of the institutional sociology of science. See Robin Blackburn, ed., *Ideology in Social Science* (London: Fontana, 1972); and Les Levidow, ed., *Radical Science Essays* (London: Free Association, 1986). For a left critique of this approach, see John O'Neill, "Marxism and the Two Sciences," *Philosophy of the Social Sciences* 11 (1981): 281–302. While this approach is certainly illuminating, I think it is much more fruitful to focus on the discursive aspects of science's text, particularly methodology. A whole host of studies have emerged in recent years that significantly extend the traditional sociology of science in this direction. See, for example, Michael Mulkay, *The Word and the World: Explorations in the Form of Sociological Analysis* (London: Allen and Unwin, 1985); G. Nigel Gilbert and Michael Mulkay, *Opening Pandora's Box: A Sociological Analysis of Scientists' Discourse* (Cambridge: Cambridge University Press, 1984); Michael Mulkay, "Action and Belief or Scientific Discourse? A Possible Way of Ending Intellectual Vassalage in Social Studies of Science," *Philosophy of the Social Sciences* 11 (1981): 163–71; Karin D. Knorr-Cetina and Michael Mulkay, eds., *Science Observed: Perspectives on the Social Study of Science* (Beverly Hills: Sage, 1983); Michael Mulkay, Jonathan Potter and S. Yearley, "Why an Analysis of Scientists' Discourse Is Needed," in Knorr-Cetina and Mulkay, eds., *Science*

Observed: Perspectives on the Social Study of Science; Karin D. Knorr(-Cetina) and D. Knorr, *From Scenes to Scripts: On the Relationship Between Research and Publication in Science,* Research memorandum 132 (Vienna: Institute for Advanced Studies, 1978); Karin D. Knorr-Cetina, *The Manufacture of Knowledge: An Essay on the Constructivist and Contextual Nature of Science* (New York: Pergamon, 1981); Bruno Latour and Steve Woolgar, *Laboratory Life: The Social Construction of Scientific Facts* (Beverly Hills: Sage, 1979); Charles Bazerman, "What Written Knowledge Does: Three Examples of Academic Discourse," *Philosophy of the Social Sciences* 11 (1981): 361–87; Steven Yearley, "Textual Persuasion: The Role of Social Accounting in the Construction of Scientific Arguments," *Philosophy of the Social Sciences* 11 (1981): 409–35. While these discourse-analytic perspectives on science offer a more illuminating view of their narratively constitutional nature, most of these studies, notably including the work of Knorr-Cetina and Mulkay, are overly ethnomethodological in nature. That is, they lack a political focus on the nature of positivist discourse as a political language game in its own right. The sociology of science, even in its discourse-analytic reformulation, is still sociology, not critical theory. I want to both build on and move beyond this discourse-analytic reformulation of the traditional institutionalist perspective on science. It is not enough to know how positivist narrative "works." It must also be worked through, hence opposed, in the way of a variety of left-wing critics of reading and writing. See, for example, Fredric Jameson, *The Political Unconscious: Narrative as a Socially Symbolic Act* (Ithaca: Cornell University Press, 1981). And see his "Ideology of the Text," *Salmagundi* 31 (1975–76): 204–46. And then see William Dowling, *Jameson, Althusser, Marx* (Ithaca: Cornell University Press, 1984). Unfortunately, these left-wing deconstructors virtually ignore science as a narrative topic. They concentrate on literature and other cultural expressions. About the only people to have examined science as a productive form of ideology are the Frankfurt School. See Herbert Marcuse, *An Essay on Liberation* (Boston: Beacon Press, 1969); Jurgen Habermas, *Knowledge and Human Interests* (Boston: Beacon Press, 1971); and my own "Marcuse and Habermas on New Science," *Polity* 14, no. 2 (Winter 1976): 158–81. Here, with the aid of Derrida, I extend the Frankfurt critique of science as ideology in a literary-theoretic direction. I side with Marcuse against Habermas in endorsing the possibilities of a "new science," sketched in my concluding chapter. See Jacques Derrida, *Of Grammatology* (Baltimore: Johns Hopkins University Press, 1976) and *Writing and Difference* (Chicago: University of Chicago Press, 1978).

4. J. Richard Udry and John O.G. Billy, "The Initiation of Coitus in Early Adolescence," *American Sociological Review* 52 (December 1987): 841–55.

5. See, for example, Abraham Kaplan, *The Conduct of Inquiry* (San Francisco: Chandler, 1964); Scott Greer, *The Logic of Social Inquiry* (Chicago: Aldine, 1969); Aaron Cicourel, *Method and Measurement in Sociology* (Glencoe, Ill.: Free Press, 1964).

6. For a typically technical discussion of statistical issues, see M. H. Birnbaum, "The Devil Rides Again: Correlation as an Index of Fit," *Psychological Bulletin* 79 (1973): 239–42.

7. The two best discussions of science as discourse are Knorr-Cetina, *The Manufacture of Knowledge;* and Gilbert and Mulkay, *Opening Pandora's Box.* (Cambridge: Cambridge University Press, 1984). Both focus on natural science. Knorr-Cetina concentrates much more than do Gilbert and Mulkay on the "conversion" of laboratory practice into journal discourse. While Knorr-Cetina comes much closer than do Gilbert and Mulkay to recognizing the political nature of science's concealment of its own authorship, both belong much more squarely in the anthropological-ethnomethodological tradition than does my study here. I read science through a Marxist version of recent literary theory. They examine exactly what happens in the laboratory, while I address what "happens" in the journals themselves.

8. See G. Nigel Gilbert, "Refencing as Persuasion," *Social Studies of Science* 7 (1977): 113–22; D.C. Anderson, "Some Organisational Features in the Local Production of a Plausible Text," *Philosophy of the Social Sciences* 8 (1978): 113–35; and J. Law and R. Williams, "Put-

ting Facts Together: A Study of Scientific Persuasion," *Social Studies of Science* 12 (1982): 535–58.

9. For a discussion of mainstream journal sociology over time, see Patricia Wilner, "The Main Drift of Sociology Between 1936 and 1984," *History of Sociology* 5, no. 2 (1985): 1–20.

10. See Bazerman, "What Written Knowledge Does"; Walter Benjamin, "The Author as Producer," in Andrew Arato and Eike Gebhardt, eds., *The Essential Frankfurt School Reader* (New York: Urizen, 1978); and Karin D. Knorr, "Producing and Reproducing Knowledge: Descriptive or Constructive? Toward a Model of Research Production," *Social Science Information* 16 (1977): 969–96.

11. For an early reading of mathematics as a social text, see Dirk J. Struik, "The Sociology of Mathematics," *Science and Society* 6 (1942): 58–70.

12. See Gilbert, "Referencing as Persuasion"; and see James Wayne Dye, "The Poetization of Science," in Mark Amsler, ed., *The Languages of Creativity* (Newark: University of Delaware Press, 1986).

13. Kingsley Davis and W.E. Moore, "Some Principles of Stratification," *American Sociological Review* 10 (1942): 242–49.

14. See Roland Barthes, "The Death of the Author," in Stephen Heath, ed., *Image-Music-Text: Roland Barthes* (London: Fontana, 1977).

15. For discussions of the relationship between text and world today, see Agger, *Fast Capitalism;* and Timothy W. Luke, *Screens of Power: Ideology, Domination and Resistance in Informational Society* (Champaign: University of Illinois Press, 1989).

16. See Jurgen Habermas, "Technology and Science as 'Ideology,'" in his *Toward a Rational Society* (Boston: Beacon Press, 1970).

17. Marcuse, *One-Dimensional Man.*

18. Ludwig Wittgenstein, *Philosophical Investigations* (Oxford: Basil Blackwell, 1953).

19. For examples of Adorno's approach to the political criticism of popular culture, see "How to Look at Television," *Quarterly of Film, Radio and Television* 8 (Spring 1954): 213–35; "A Social Critique of Radio Music," *Kenyon Review* 8, no. 2 (Spring 1945): 208–17; and "The Stars Down to Earth: The *Los Angeles Times* Astrology Column: A Study in Secondary Synthesis," *Telos* 19 (Spring 1974): 13–90.

20. For different but complementary discussions of discipline in a positivist world, see Michel Foucault, *Discipline and Punish: The Birth of the Prison* (London: Allen Lane, 1979); and Agger *Socio(onto)logy.*

21. Russell Jacoby, "A Falling Rate of Intelligence?" *Telos 27* (1976): 141–46.

22. I read Marx's reading of capitalism in my *Fast Capitalism,* notably chapter 3, "Books and Things: Texts in a Material World."

23. See Norman Stockman, *Antipositivist Theories of the Sciences* (Dordrecht: D. Reidel, 1984).

24. For a discussion of how what science "really" does should inform the philosophy of science, see Roy Bhaskar, *A Realist Theory of Science* (Sussex: Harvester, 1978); also see Knorr-Cetina, *The Manufacture of Knowledge.*

25. See G. Nigel Gilbert and Michael Mulkay, "Scientists' Theory Talk," *Canadian Journal of Sociology* 8 (1983): 179–97.

26. For discussions of the constitutional nature of citation, see D. Sullivan, H.D. White, and E.J. Barboni, "Co-Citation Analysis of Science: An Evaluation," *Social Studies of Science* 7 (1977): 223–40. Also see Nicholas Mullins, L. Hargens, P. K. Hecht, and E. L. Kick, "The Group Structure of Co-Citation Clusters: A Comparative Study," *American Sociological Review* 42 (1977): 552–62.

27. For general discussions of ethnomethodology, see Harold Garfinkel, *Studies in Ethnomethodology* (Englewood Cliffs, N.J.: Prentice-Hall, 1967); and Hugh Mehan and Houston Wood, *The Reality of Ethnomethodology* (New York: Wiley, 1975). For a more

focused discussion of science from this perspective, see Michael Mulkay, *Science and the Sociology of Knowledge* (London: Allen and Unwin, 1979).

28. See Robert A. Day, *How to Write and Publish a Scientific Paper* (Philadelphia: ISI Press, 1979).

29. See John O'Neill, "The Literary Production of Natural and Social Science Inquiry," *Canadian Journal of Sociology* 6 (1981): 105–20.

30. For some pertinent remarks about the academicization of intellectual life, see Jacoby, *The Last Intellectuals*.

31. For examples of the literary criticism of criticism, see Harold Bloom, *A Map of Misreading* (New York: Oxford University Press, 1975); and Gerald Graff, *Literature Against Itself: Literary Ideas in Modern Society* (Chicago: University of Chicago Press, 1977). For an interrogation of the nature of literature, see Jean-Paul Sartre, *What Is Literature?* (New York: Harper & Row, 1965).

32. Knorr-Cetina offers an "anthropological" account of science in *The Production of Knowledge*. See especially 17–22.

33. Axel van den Berg, "Critical Theory: Is There Still Hope?", *American Journal of Sociology* 86, no. 3 (1980): 449–478.

34. See L.J. Anthony, H. East, and M. Slater, "The Growth of Literature in Physics," *Reports on Progress in Physics* 32 (1969): 709–67.

35. See R. Sharratt, *Reading Relations: Structures of Literary Production* (Brighton: Harvester, 1982).

36. For a discussion of the politics of method, see John J. Sewart, "Critical Theory and the Critique of Conservative Method," *American Sociologist* 13 (1978): 15–22.

37. For a rather ironic example of authors who study mainstream journal science using mainstream method, see Linda Grant, Kathryn B. Ward, and Xue Lan Rong, "Is There an Association Between Gender and Methods in Sociological Research?" *American Sociological Review* 52 (December 1987): 856–62.

38. I explore this theme at greater length in my *Fast Capitalism*.

39. For important discussions of the politics of communication, see Basil Bernstein, *Class, Codes and Control* (London: Routledge and Kegan Paul, 1971); and Claus Mueller, *The Politics of Communication* (New York: Oxford University Press, 1973).

40. For a general discussion of the politics of academic writing, see Linda Brodkey, *Academic Writing as Social Practice* (Philadelphia: Temple University Press, 1987). For a discussion particularly about psychology, see Jonathan Potter, P. Stringer, and M. Wetherell, *Social Texts and Contexts: Literature and Social Psychology* (London: Routledge and Kegan Paul, 1983).

2: Scripting Causality

1. See Wolfgang Iser, *The Implied Reader: Patterns of Communication in Prose Fiction from Bunyan to Beckett* (Baltimore: Johns Hopkins University Press, 1974) and *The Act of Reading: A Theory of Aesthetic Response* (Baltimore: Johns Hopkins University Press, 1978).

2. This is Wittgenstein's well-known notion of knowledge as a "language game," taken from his *Philosophical Investigations*.

3. For a discussion of the relevance of notions such as paradigm and literature for the philosophy of science, see Margaret Masterman, "The Nature of a Paradigm," in Imre Lakatos and Alain Musgrove, eds., *Criticism and the Growth of Knowledge* (Cambridge: Cambridge University Press, 1970).

4. For a discussion of the "two sciences," see Knorr-Cetina, *The Manufacture of Knowledge*, especially 136–51.

5. For a critique of social-science positivism, see the Institute for Social Research, *Aspects of Sociology* (Boston: Beacon Press, 1972). See especially the chapter "The Concept of Sociology."

6. For a discussion of the use of figure in science, see Gilbert and Mulkay, *Opening Pandora's Box*, especially the chapter "Working Conceptual Hallucinations."

7. For a critique of the crisis of Western culture focusing on issues of science and mathematics, see Edmund Husserl, *The Crisis of the European Sciences and Transcendental Phenomenology* (Evanston: Northwestern University Press, 1970). For thoughts on mathematics from a philosophical point of view, see Ludwig Wittgenstein, *Remarks on the Foundations of Mathematics* (Cambridge: MIT Press, 1978).

8. For a selection of papers on mathematical sociology, see Thomas J. Fararo, *Mathematical Ideas and Sociological Theory: A Special Issue of the Journal of Mathematical Sociology* (New York: Gordon and Breach, 1984).

9. For example, see Anatol Rapoport, *Mathematical Models in the Social and Behavioral Sciences* (New York: Wiley, 1983).

10. See Kenneth C. Land, "Mathematical Formalization of Durkheim's Theory of Division of Labor," in E. Borgatta and G. Bohrnstedt, eds., *Sociological Methodology* (San Francisco: Jossey-Bass, 1970).

11. See Michael Lynch, *Art and Artifact in Laboratory Science* (London: Routledge and Kegan Paul, 1981).

12. See Ernest Nagel, *The Structure of Science* (New York: Harcourt, Brace and World, 1961); and Carl G. Hempel, *Philosophy of Natural Science* (Englewood Cliffs, N.J.: Prentice-Hall, 1966).

13. See A.W. McHoul, *Telling How Texts Talk* (London: Routledge and Kegan Paul, 1982).

14. Anthony, East, and Slater, "The Growth of Literature in Physics."

15. See Gilbert, "Referencing as Persuasion"; and G. Nigel Gilbert, "The Transformation of Research Findings into Scientific Knowledge," *Social Studies of Science* 6 (1976): 281–306.

16. See Mullins, Hargens, Hecht, and Kick, "The Group Structure of Co-Citation Clusters."

17. See Michael Mulkay and G. Nigel Gilbert, "Accounting for Error: How Scientists Construct Their Social World When They Account for Correct and Incorrect Belief," *Sociology* 16 (1982): 165–83.

18. For discussions of the concealed metaphysics of positivist epistemology, see Derek Phillips, *Knowledge From What? Theories and Methods in Social Research* (Chicago: Rand McNally, 1971); and Joseph Gusfield, "The Literary Rhetoric of Science," *American Sociological Review* 41 (1976): 16–34.

19. See Peter Rabinowitz, "Truth in Fiction: A Reexamination of Audiences," *Critical Inquiry* 4 (1977): 121–42.

20. Iser makes this point in the development of his reader-response theory. See his *The Act of Reading.*

21. For example, see John Hagan, A.R. Gillis, and John Simpson, "The Class Structure of Gender and Delinquency: Toward a Power-Control Theory of Common Delinquent Behavior," *American Journal of Sociology* 90, no. 6 (1985): 1151–75.

22. See the special issue edited by Robert M. Maniquis on "Marxists and the University," *Humanities in Society* 6, nos. 2–3 (Spring and Summer 1983).

23. I subscribe to the critique of positivism essayed by Max Horkheimer and Theodor W. Adorno in their *Dialectic of Enlightenment* (New York: Herder and Herder, 1972). See especially the chapter "The Concept of Enlightenment."

24. See Richard Rorty, "Philosophy as a Kind of Writing: An Essay on Derrida," *New Literary History* 10 (1978): 141–60.

25. J.L. Austin, *How to Do Things with Words* (Oxford: Oxford University Press, 1962).

26. See Ben Agger, "A Critical Theory of Dialogue," *Humanities in Society* 4, no. 1 (Winter 1981): 7–30.

27. Roland Barthes, *Writing Degree Zero* (New York: Hill and Wang, 1967).

28. Knorr-Cetina and Knorr, From *Scenes to Scripts*.

29. See Friedrich A. Hayek, *The Road to Serfdom* (Chicago: University of Chicago Press, 1944).

3: The First Page

1. See D. Silverman and B. Torode, *The Material Word* (London: Routledge and Kegan Paul, 1980).

2. See Mulkay, "Action and Belief or Scientific Discourse?"

3. See Merton, *The Sociology of Science*.

4. For a perspective on the literary nature of the human sciences, see Paul Ricoeur, *Hermeneutics and the Human Sciences* (Cambridge: Cambridge University Press, 1981).

5. For a discussion of this, see Knorr-Cetina, *Manufacture of Knowledge*, 101–7.

6. Lawrence E. Cohen and Kenneth C. Land, "Age Structure and Crime: Symmetry Versus Asymmetry and the Projection of Crime Rates Through the 1990s," *American Sociological Review* 52 (April 1987): 170.

7. J. Craig Jenkins and Craig M. Eckert, "Channeling Black Insurgency: Elite Patronage and Professional Social Movement Organizations in the Development of the Black Movement," *American Sociological Review* 51 (December 1986): 812.

8. Donald Palmer, Roger Friedland and Jitendra V. Singh, "The Ties That Bind: Organizational and Class Bases of Stability in a Corporate Interlock Network," *American Sociological Review* 51 (December 1986): 781.

9. Hagan, Gillis and Simpson, "The Class Structure of Gender and Delinquency," 1151.

10. Karen A. Miller, Melvin L. Kohn, and Carmi Schooler, "Educational Self-Direction and Personality," *American Sociological Review* 51 (June 1986): 372.

11. Karin D. Knorr-Cetina devotes an instructive chapter of her *Manufacture of Knowledge* to a discussion of the "conversion" of laboratory results into written discourse. See the chapter "The Scientist as a Literary Reasoner, or the Transformation of Laboratory Reason."

12. See Pierre Bourdieu, "The Specificity of the Scientific Field and the Social Conditions of the Progress of Reason," *Social Science Information* 14, no. 6 (1975): 19–47.

13. See Ben Agger, "Marxism, Feminism, Deconstruction," in Christine Gailey and Viana Muller, eds., *Dialectical Anthropology: The Critical Ethnology of Stanley Diamond* (forthcoming).

14. For a discussion of the nature of the referee process used in mainstream academic journal publishing, see Von Bakanic, Clark McPhail, and Rita J. Simon, "The Manuscript Review and Decision-Making Process," *American Sociological Review* 52 (October 1987): 631–42; Lowell L. Hargens, "Scholarly Consensus and Journal Rejection Rates," *American Sociological Review* 53 (February 1988): 139–51; Stephen Cole, Gary Simon, and Jonathan R. Cole, "Do Journal Rejection Rates Index Consensus?" *American Sociological Review* 53 (February 1988): 152–56; Lowell L. Hargens, "Further Evidence of Field Differences in Consensus from the NSF Peer Review Studies," *American Sociological Review* 53 (February 1988): 157–60.

4: Authorizing Authority: The Literary Irony of Science

1. Marta Tienda, Shelley A. Smith, and Vilma Ortiz, "Industrial Restructuring, Gender Segregation, and Sex Differences in Earnings," *American Sociological Review* 52 (April 1987): 196.

2. York W. Bradshaw, "Urbanization and Underdevelopment: A Global Study of Modernization, Urban Bias, and Economic Dependency," *American Sociological Review* 52 (April 1987): 225.

3. Hagan, Gillis, and Simpson, "The Class Structure of Gender and Delinquency," 1158.

4. Bruce London, "Structural Determinants of Third World Urban Change: An Ecological and Political Analysis," *American Sociological Review* 52 (February 1987): 31.

5. For an exploration of this issue, see Peter Winch, *The Idea of a Social Science* (London: Routledge and Kegan Paul, 1969).

6. Sally S. Simpson, "The Decomposition of Antitrust: Testing a Multi-Level, Longitudinal Model of Profit-Squeeze," *American Sociological Review* 51 (December 1986): 860.

7. Ibid.

8. Robert V. Robinson and Maurice A. Garnier, "Class Reproduction Among Men and Women in France: Reproduction Theory on Its Home Ground," *American Journal of Sociology* 91, no. 2 (1985): 257.

9. Joan R. Kahn and William M. Mason, "Political Alienation, Cohort Size, and the Easterlin Hypothesis," *American Sociological Review* 52 (April 1987): 156.

10. Hagan, Gillis, and Simpson, "The Class Structure of Gender and Delinquency," 1152.

11. Ibid., 1151.

12. Frederick D. Weil, "Cohorts, Regimes, and the Legitimation of Democracy: West Germany Since 1945," *American Sociological Review* 52 (June 1987): 310.

13. Cohen and Land, "Age Structure and Crime," 170.

14. Robert D. Mare and Meichu D. Chen, "Further Evidence on Sibship Size and Educational Stratification," *American Sociological Review* 51 (June 1986): 403.

15. Cohen and Land, "Age Structure and Crime," 180.

16. Palmer, Friedland, and Singh, "The Ties That Bind," 782.

17. Bradshaw, "Urbanization and Underdevelopment," 225.

18. I address this issue of disciplinary privilege in the last chapter of my *Socio(onto)logy*, "Open Text, Open World."

19. Cohen and Land, "Age Structure and Crime," 171–72.

20. Amy S. Wharton and James N. Baron, "So Happy Together? The Impact of Gender Segregation on Men at Work," *American Sociological Review* 52 (October 1987): 574.

21. Miller, Kohn, and Schooler, "Educational Self-Direction and Personality," 372.

22. John R. Sutton, "Doing Time: Dynamics of Imprisonment in the Reformist State," *American Sociological Review* 52 (October 1987): 612.

23. Ibid., 613–4.

24. Knorr-Cetina discusses this issue in her *Manufacture of Knowledge*, 129.

25. Knorr-Cetina addresses this issue in her *Manufacture* in the chapter "The Scientist as an Indexical Reasoner: The Contextuality and the Opportunism of Research." She suggests that topic choice is often governed opportunistically by the unpredictable contingencies of the laboratory routine.

26. Bradshaw, "Urbanization and Underdevelopment," 224.

27. Hagan, Gillis, and Simpson, "The Class Structure of Gender and Delinquency," 1152.

28. Jonathan Turner, "Toward a Sociological Theory of Motivation," *American Sociological Review* 52 (February 1987): 16.

29. Miller, Kohn, and Schooler, "Educational Self-Direction and Personality," 372.

30. Davis and Moore, "Some Principles of Stratification."

5: Science Aura I: Method as Topic

1. For a discussion of this, see Gilbert and Mulkay, *Opening Pandora's Box*, especially 162–67.

2. See Wittgenstein, *Remarks on the Foundations of Mathematics.*

3. Knorr-Cetina, in her *Manufacture of Knowledge,* calls the scientist a "literary reasoner." See her chapter "The Scientist as a Literary Reasoner, Or the Transformation of Laboratory Reason."

4. For a discussion of this particular line of argument with respect to the alleged maleness of science (and thus the appropriateness of a feminist methodology), see Evelyn Fox Keller, *Reflections on Gender and Science* (New Haven: Yale University Press, 1985).

5. For discussions of the constitutional nature of methodology, see Cicourel, *Method and Measurement in Sociology;* and Phillips, *Knowledge From What?*

6. See Agger, "Marcuse and Habermas on New Science."

7. See Habermas, "Technology and Science as 'Ideology,'" in his *Toward a Rational Society.*

8. For an early (1923) discussion of reification, see Lukacs, *History and Class Consciousness.* For an update of sorts, see Horkheimer and Adorno, *Dialectic of Enlightenment.* For a discussion of positivism as a disciplining literary strategy, see Agger, *Socio(onto)logy,* especially the chapter "Social Fact, Social Fate, Social Text."

9. Robinson and Garnier, "Class Reproduction Among Men and Women in France," 260.

10. Jenkins and Eckert, "Channeling Black Insurgency," 818.

11. Grant, Ward, and Rong, "Is There an Association Between Gender and Methods in Sociological Research?" 858.

12. Wharton and Baron, "So Happy Together?" 574.

13. Michael Hout and Andrew M. Greeley, "The Center Doesn't Hold: Church Attendance in the United States, 1940–1984," *American Sociological Review* 52 (June 1987): 329.

14. See Knorr-Cetina's discussion of "conversion" in her *Manufacture of Knowledge,* 95, 131, 152.

15. Melvin E. Thomas and Michael Hughes, "The Continuing Significance of Race: A Study of Race, Class, and Quality of Life in America, 1972–1985," *American Sociological Review* 51 (December 1986): 834.

16. Hagan, Gillis, and "The Class Structure of Gender and Delinquency," 1157

17. Maureen T. Hallinan and Richard A. Williams, "The Stability of Students' Interracial Friendships," *American Sociological Review* 52 (October 1987): 655–56.

18. See Karin Knorr-Cetina, "Tinkering Toward Success: Prelude to a Theory of Scientific Practice," *Theory and Society* 8 (1979): 347–76.

19. I discuss science as a form of art in my "On Happiness and the Damaged Life," in John O'Neill, ed., *On Critical Theory* (New York: Seabury, 1976).

20. For an ethnomethodological perspective on validity, see Peter McHugh et al., *On the Beginning of Social Inquiry* (London: Routledge and Kegan Paul, 1974).

21. Wharton and Baron, "So Happy Together?" 576.

22. Turner, "Toward a Sociological Theory of Motivation," 26.

23. Ibid., 23.

24. Ibid., 26.

25. Simpson, "The Decomposition of Antitrust," 861.

26. Edgar Kiser and K.A. Drass, "Changes in the Core of the World System and the Production of Utopian Literature," *American Sociological Review* 52 (April 1987): 289.

27. London, "Structural Determinants of Third World Urban Change," 40.

28. Simpson, "The Decomposition of Antitrust," 865.

29. Ibid., 864.

30. Sutton, "Doing Time," 617.

31. For a discussion of media from this point of view, see Luke, *Screens of Power.*

32. For discussions of the concept of totality, see Lukacs, *History and Class Consciousness,* especially the chapter "Reification and the Consciousness of the Proletariat"; also see Martin Jay, *Marxism and Totality* (Berkeley: University of California Press, 1984).

33. See Karl Popper, *The Logic of Scientific Discovery* (New York: Science Edition, 1961).

34. See Hans Zetterberg, *On Theory and Verification in Sociology* (Totowa, N.J.: Bedminister, 1955).

35. See S. Woolgar, "Discovery: Logic and Sequence in a Scientific Text," in K. Knorr, R. Krohn, and R. Whitley, eds., *The Social Process of Scientific Investigation, Sociology of the Sciences,* Yearbook Vol. 4 (Dordrecht: D. Reidel, 1980).

36. Thomas and Hughes, "The Continuing Significance of Race," 835.

37. Paul William Kingston and Steven L. Nock, "Time Together Among Dual-Earner Couples," *American Sociological Review* 52 (June 1987): 395–96.

38. London, "Structural Determinants of Third World Urban Change," 40.

39. Bradshaw, "Urbanization and Underdevelopment," 228.

40. There are some exceptions. See Hilary Rose and Steven Rose, eds., *The Radicalization of Science: Ideology of/in the Natural Sciences* (London: Macmillan, 1976). Also see Levidow, ed., *Radical Science Essays.*

41. Courtney Brown, "Voter Mobilization and Party Competition in a Volatile Electorate," *American Sociological Review* 52 (February 1987): 62.

42. Miller, L. Kohn, and Schooler, "Educational Self-Direction and Personality," 374.

43. Mare and Chen, "Further Evidence on Sibship Size and Educational Stratification," 407.

44. Robinson and Garnier, "Class Reproduction Among Men and Women in France," 260.

45. Udry and Billy, "Initiation of Coitus in Early Adolescence," 844.

46. Hallinan and Williams, "The Stability of Students' Interracial Friendships," 659–60.

47. Karen A. Miller, Melvin L. Kohn, and Carmi Schooler, "Educational Self-Direction and Personality," *American Sociological Review* 51 (June 1986): 377.

48. Robert L. Kaufman and Paul G. Schervish, "Using Adjusted Crosstabulations to Interpret Log-Linear Relationships," *American Sociological Review* 51 (October 1986): 717.

49. Bradshaw, "Urbanization and Underdevelopment," 228.

50. Hout and Greeley, "The Center Doesn't Hold," 333.

51. Turner, "Toward a Sociological Theory of Motivation," 26.

52. See Greer, *The Logic of Social Inquiry.*

53. See Garfinkel, *Studies in Ethnomethodology.*

6: Science Aura II: Findings Found

1. Michael E. Sobel, Michael Hout, and Otis Dudley Duncan, "Exchange, Structure, and Symmetry in Occupational Mobility," *American Journal of Sociology* 91, no. 2 (1985): 626.

2. Lynn Smith-Lovin, John V. Skvoretz, and Charlotte G. Hudson, "Status and Participation in Six-Person Groups: A Test of Skovertz's Comparative Status Model," *Social Forces* 64, no. 4 (1986): 1000.

3. Wharton and Baron, "So Happy Together?" 583.

4. Tienda, Smith, and Ortiz, "Industrial Restructuring, Gender Segregation, and Sex Differences in Earnings," 199–200.

5. Cohen and Land, "Age Structure and Crime," 177–78.

6. Hout and Greeley, "The Center Doesn't Hold," 338.

7. Cohen and Land, "Age Structure and Crime," 179.

8. See C. A. Lave and J. G. March, *An Introduction to Models in the Social Sciences* (New York: Harper & Row, 1975); also see Mark Amsler, ed., *The Languages of Creativity: Models, Problem-Solving, Discourse* (Newark: University of Delaware Press, 1986).

9. Mare and Chen, "Further Evidence on Sibship Size and Educational Stratification," 406.

10. London, "Structural Determinants of Third World Urban Change," 39–40.

11. Bradshaw, "Urbanization and Underdevelopment," 232.

12. Palmer, Friedland, and Singh, "The Ties That Bind," 794.

13. For example, see Patricia Willner, "The Main Drift of Sociology Between 1936 and 1984." Also see Bazerman, "What Written Knowledge Does."

14. See O'Neill, "Marxism and the Two Sciences."

15. Messner and South, "Economic Deprivation, Opportunity Structure, and Robbery Victimization," 987.

16. Charles N. Halaby, "Worker Attachment and Workplace Authority," *American Sociological Review* 51 (October 1986): 645.

17. See Theodor W. Adorno, *The Jargon of Authenticity* (Evanston: Northwestern University Press, 1970).

18. See Horkheimer, *Eclipse of Reason;* and see Ben Agger, "Dialectical Sensibility I: Critical Theory, Scientism and Empiricism," *Canadian Journal of Political and Social Theory* 1, no. 1 (1977): 3–34; and Ben Agger, "Dialectical Sensibility II: Towards a New Intellectuality," *Canadian Journal of Political and Social Theory* 1, no. 2 (1977): 47–57.

19. Cohen and Land, "Age Structure and Crime," 174.

20. Weil, "Cohorts, Regimes, and the Legitimation of Democracy," 316.

21. Sobel, Hout, and Duncan, "Exchange, Structure, and Symmetry in Occupational Mobility," 371.

22. For a discussion of this basic semiotic principle, see Umberto Eco, *The Role of the Reader: Explanation in the Semiotics of Texts* (Bloomington: Indiana University Press, 1979).

23. I develop this theme of booklessness in my *Fast Capitalism.*

24. Brown, "Voter Mobilization and Party Competition in a Volatile Electorate," 65.

25. Ibid., 68.

26. Bradshaw, "Urbanization and Underdevelopment," 229.

27. For a discussion of science as alienated mental labor, see Alfred Sohn-Rethel, "Science as Alienated Consciousness," *Radical Science Journal* 2/3 (1975): 72–101.

28. See Jacques Derrida, *Writing and Difference.*

29. Theodor W. Adorno, *Negative Dialectics* (New York: Seabury, 1973).

30. Jurgen Habermas, *The Theory of Communicative Action,* vols. 1 and 2 (Boston: Beacon Press, 1984 and 1987).

31. Michel Foucault, *The Archaelogy of Knowledge* (New York: Harper & Row, 1976); and *Discipline and Punish: The Birth of the Prison* (London: Allen Lane, 1979).

32. See Paul Feyerabend, *Against Method* (London: New Left Books, 1975).

33. See Geoffrey Hartman, *Saving the Text: Literature, Derrida, Philosophy* (Baltimore: Johns Hopkins University Press, 1981).

7: Science Aura III: Findings Figured

1. Habermas stands on one side of this debate and Marcuse on the other. See Habermas, *Knowledge and Human Interests;* and Marcuse, *An Essay on Liberation.* See Agger, "Marcuse and Habermas on New Science."

2. Susan Sontag, *On Photography* (New York: Farrar, Straus and Giroux, 1977).

3. Jacoby, "A Falling Rate of Intelligence?"

4. See Josue V. Harari, ed., *Textual Strategies* (Ithaca: Cornell University Press, 1979).

5. See Ryan, *Marxism and Deconstruction;* also see Terry Eagleton *Marxism and Literary Criticism* (London: Methuen, 1976), and "Marxism, Structuralism and Poststructuralism," *Diacritics* 15, no. 4 (Winter 1985): 2–56.

6. See Alex Callinicos, "Postmodernism, Post-structuralism, Post-Marxism?" *Theory, Culture and Society* 2, no. 3 (1985): 85–101; Andrew Parker, "Of Politics and Limits: Derrida Re-Marx," *SCE Reports* 8 (1980): 83–104; Nancy Fraser, "The French Derrideans: Politicizing Deconstruction or Deconstructing the Political?" *New German Critique* 33 (1984): 127–54.

7. Gilbert and Mulkay, in their *Opening Pandora's Box,* count eight features of figural expression in science. See 145–48.

8. For a discussion of validity, see E.D. Hirsch, *Validity in Interpretation* (New Haven: Yale University Press, 1967).

9. For a different version of this discussion, see Ben Agger and Allan Rachlin, "Left-Wing Scholarship: Current Contradictions of Academic Production," *Humanities in Society* 6, nos. 2-3 (1983): 241–56.

10. For a discussion of different modes of theorizing, see Cary Nelson, ed., *Theory in the Classroom* (Champaign: University of Illinois Press, 1986).

11. In her study of the laboratory and literary production of natural science, *The Manufacture of Knowledge,* Knorr-Cetina interestingly notes the opposite tendency in natural science. Writers are compelled to shorten their methodology sections. In fact, they may be encouraged to shorten only prose methodology but not figural methodology. In any event, one can understand this in terms of the relative immaturity of the social sciences with respect to a natural-science model of scientificity.

12. For a discussion of the philosophical foundations of natural science by the father of the famous principle of "indeterminacy," see Werner Heisenberg, *Philosophical Problems of Quantum Physics* (Woodbridge, Conn.: Ox Bow Press, 1979).

13. Susan Sontag, *On Photography* (New York: Farrar, Straus and Giroux, 1977).

8: Concluding Science

1. Halaby, "Worker Attachment and Workplace Authority," 646.

2. Turner, "Toward a Sociological Theory of Motivation," 26.

3. Claude Levi-Strauss, *The Savage Mind* (Chicago: University of Chicago Press, 1966), especially 16–36.

4. Mare and Chen, "Further Evidence on Sibship Size and Educational Stratification," 412.

5. Hout and Greeley, "The Center Doesn't Hold," 341.

6. Bruce G. Link, "Understanding Labeling Effects in the Area of Mental Disorders: An Assessment of the Effects of Expectations of Rejection," *American Sociological Review* 52 (February 1987): 109.

7. Udry and Billy, "Initiation of Coitus in Early Adolescence," 852.

8. Grant, Ward, and Rong, "Is There an Association Between Gender and Methods in Sociological Research?" 861.

9. Wharton and Baron, "So Happy Together?" 585.

10. Ibid., 584.

11. Miller, Kohn, and Schooler, "Educational Self-Direction and Personality," 388.

12. Ibid., 388.

13. Kingston and Nock, "Time Together Among Dual-Earner Couples," 399.

14. Hagan, Gillis, and Simpson, "The Class Structure of Gender and Delinquency," 1175.

15. Miller, Kohn, and Schooler, "Educational Self-Direction and Personality," 388.

16. Wharton and Baron, "So Happy Together?" 585.

17. Udry and Billy, "Initiation of Coitus in Early Adolescence," 852.

18. Wharton and Baron, "So Happy Together?" 586.

19. Smith-Lovin, Skvoretz, and Hudson, "Status and Participation in Six- Person Groups," 1003.

20. Wharton and Baron, "So Happy Together?" 586.

21. Messner and South, "Economic Deprivation, Opportunity Structure, and Robbery Victimization," 989.

22. Cohen and Land, "Age Structure and Crime," 179.

23. Hagan, Gillis, and Simpson, "The Class Structure of Gender and Delinquency," 1175.

24. Robinson and Garnier, "Class Reproduction Among Men and Women in France," 278.

25. Turner, "Toward a Sociological Theory of Motivation," 26.

26. Mare and Chen, "Further Evidence on Sibship Size and Educational Stratification," 412.

27. Hallinan and Williams, "The Stability of Students' Interracial Friendships," 663.

28. Tienda, Smith, and Ortiz, "Industrial Restructuring, Gender Segregation, and Sex Differences in Earnings," 208.

9: Journal Theory

1. James S. Coleman, "Social Theory, Social Research, and a Theory of Action," *American Journal of Sociology* 91, no. 6 (1986): 1309.

2. Paul DiMaggio, "Classification in Art," *American Sociological Review* 52 (August 1987): 441.

3. David Sciulli, "Voluntaristic Action as a Distinct Concept: Theoretical Foundations of Societal Constitutionalism," *American Sociological Review* 51 (December 1986): 745.

4. See Nicholas Mullins, *Theories and Theory Groups in Contemporary American Sociology* (New York: Harper & Row, 1973).

5. See, for example, Jere Cohen, Lawrence Hazelrigg, and Whitney Pope, "DeParsonizing Weber: A Critique of Parsons' Interpretation of Weber's Sociology," *American Sociological Review* 40 (1975): 229-41; Whitney Pope, "Classic on Classic: Parsons' Interpretation of Durkheim," *American Sociological Review* 38 (1973): 399-415; and Whitney Pope, Jere Cohen, and Lawrence E. Hazelrigg, "On the Divergence of Weber and Durkheim: A Critique of Parsons' Convergence Thesis," *American Sociological Review* 40 (1975): 417-27.

6. Charles Camic, "The Making of a Method: A Historical Reinterpretation of the Early Parsons," *American Sociological Review* 52 (August 1987): 436.

7. Ibid., 424.

8. Sciulli, "Voluntaristic Action as a Distinct Concept," 745.

9. Camic, "The Making of a Method," 423.

10. Ibid., 422.

11. Sciulli, "Voluntaristic Action as a Distinct Concept," 743.

12. For a discussion of how sociological theorists made sociology, see Donald Martindale, *The Nature and Types of Sociological Theory,* 2d ed. (Boston: Houghton Mifflin, 1981).

13. The master adjudicator in mainstream sociology is George Ritzer. See his *Sociology: A Multiple Paradigm Science,* Rev. ed. (Boston: Allyn and Bacon, 1980); and *Toward an Integrated Sociological Paradigm: The Search for an Exemplar and an Image of the Subject Matter* (Boston: Allyn and Bacon, 1981).

14. Sciulli, "Voluntaristic Action as a Distinct Concept," 743.

15. Camic, "The Making of a Method," 424.

16. Sciulli, "Voluntaristic Action as a Distinct Concept," 744.

17. Ibid.

18. Ibid., 753–54.

19. Camic, "The Making of a Method," 421.

20. Ibid., 428.

21. Ibid., 430.
22. Ibid., 428.
23. Sciulli, "Voluntaristic Action as a Distinct Concept," 750.
24. Ibid., 743.
25. Ibid., 760.
26. Bradshaw, "Urbanization and Underdevelopment," 237.
27. Turner, "Toward a Sociological Theory of Motivation," 16.
28. M. Gottdiener, "Hegemony and Mass Culture: A Semiotic Approach," *American Journal of Sociology* 90, no. 5 (1985): 981.
29. Turner, "Toward a Sociological Theory of Motivation," 22–23.
30. Hagan, Gillis, and Simpson, "The Class Structure of Gender and Delinquency," 1158.
31. Adorno makes this case in his *Negative Dialectics.*
32. Camic, "The Making of a Method," 436.
33. Barney Glaser and Anselm Strauss, *The Discovery of Grounded Theory* (Chicago: Aldine, 1967).
34. Robert K. Merton, "On Sociological Theories of the Middle Range," in Merton, *On Theoretical Sociology* (New York: Free Press, 1967).
35. Bradshaw, "Urbanization and Underdevelopment," 237.
36. Ibid., 236.
37. Land has tried to do precisely this. See his "Mathematical Formalization of Durkheim's Theory of Division of Labor."
38. Sciulli, "Voluntaristic Action as a Distinct Concept," 743.
39. Ibid., 760.
40. Ibid., 761.
41. See Zetterberg, *On Theory and Verification in Sociology.*
42. James S. Coleman, "Social Theory, Social Research, and a Theory of Action," 1332.
43. Ibid., 1321–22.
44. Ibid., 1330.
45. Turner, "Toward a Sociological Theory of Motivation," 15.
46. DiMaggio, "Classification in Art," 452.
47. Ibid., 443.
48. Ibid., 443.
49. Ibid., 444.
50. Ibid., 444.
51. Ibid., 444.
52. Jenkins and Eckert, "Channeling Black Insurgency," 813.

10: Essaying a Different Text

1. Feyerabend makes this point in his *Against Method.*
2. I develop this notion of good science in the last chapter ("Open Text, Open World") of my *Socio(onto)logy.*
3. Habermas preserves this notion of the value of the Enlightenment—of reason—in his *Theory of Communicative Action.*
4. See Agger, "A Critical Theory of Dialogue."
5. Horkheimer, *Eclipse of Reason.*
6. For an extended discussion of Montaigne's concept and practice of essay, see John O'Neill, *Essaying Montaigne: A Study of the Renaissance Institution of Writing and Reading* (London: Routledge and Kegan Paul, 1982).

7. See Rorty, "Philosophy as a Kind of Writing."

8. See Jameson, *Political Unconscious.*

9. See John Gardner, *On Becoming a Novelist* (New York: Harper & Row, 1983).

10. See Sartre, *What Is Literature?;* and Walter Benjamin, "The Author as Producer," in Andrew Arato and Eike Gebhardt, eds., *The Essential Frankfurt School Reader* (New York: Urizen, 1978).

11. See Albrecht W. Wellmer, "Communications and Emancipation: Reflections on the Linguistic Turn in Critical Theory," in John O'Neill, ed., *On Critical Theory* (New York: Seabury, 1976).

12. See Walter Cohen and Peter Hohendahl, "Marxist Literary Critics: Problems and Proposals," *Humanities in Society* 6, nos. 2–3 (Spring-Summer 1983): 161–77.

13. See Jacoby, *The Last Intellectuals.*

14. See Ben Agger, "On Science as Domination," in Alkis Kontos, ed., *Domination* (Toronto: University of Toronto Press, 1975).

15. For materialist perspectives on the humanities and culture, see Cary Nelson and Lawrence Grossberg, eds., *Marxism and the Interpretation of Culture* (Champaign: University of Illinois Press, 1988).

16. For a version of this discussion, see John O'Neill, "Decolonization and the Ideal Speech Community: Some Issues in the Theory and Practice of Communicative Competence," in John Forester, ed., *Critical Theory and Public Life* (Cambridge: MIT Press, 1985).

17. See Alvin Gouldner, *The Two Marxisms: Contradictions and Anomalies in the Development of Theory* (New York: Seabury, 1980).

18. See Paul deMan, *Allegories of Reading* (New Haven: Yale University Press, 1979).

19. For a discussion of the politics of communication, see Selya Benhabib, *Critique, Norm and Utopia* (New York: Columbia University Press, 1987).

20. For a discussion of postmodernism, see Andreas Huyssen, "Mapping the Postmodern," *New German Critique* 33 (1984): 5–52.

Bibliography

Aaronson, S. "Style in Scientific Writing." *Current Contents, Life Sciences* 20 (1977): 6–15.

Adorno, Theodor W. "How to Look at Television." *Quarterly of Film, Radio and Television* 8 (Spring 1954): 213–35.

__________. *The Jargon of Authenticity.* Evanston: Northwestern University Press, 1970.

__________. *Negative Dialectics.* New York: Seabury, 1973.

__________. "A Social Critique of Radio Music." *Kenyon Review* 8, no. 2 (Spring 1945): 208–217.

__________. "The Sociology of Knowledge and its Consciousness." In Andrew Arato and Eike Gebhardt, eds., *The Essential Frankfurt School Reader,* 452–65. New York: Urizen, 1978.

__________. "The Stars Down to Earth: *The Los Angeles Times* Astrology Column: A Study in Secondary Synthesis." *Telos* 19 (Spring 1974): 13–90.

Adorno Theodor W., Hans Albert, Ralf Dahrendorf, Jurgen Habermas, Harold Pilot, and Karl R. Popper. *The Positivist Dispute in German Sociology.* London: Heinemann, 1976.

Agger, Ben. "A Critical Theory of Dialogue." *Humanities in Society* 4, no. 1 (Winter 1981): 7–30.

__________. "Dialectical Sensibility I: Critical Theory, Scientism and Empiricism." *Canadian Journal of Political and Social Theory* 1, no. 1 (1977a): 3–34.

__________. "Dialectical Sensibility II: Towards a New Intellectuality." *Canadian Journal of Political and Social Theory* 1, no. 2 (1977b): 47–57.

__________. *Fast Capitalism: A Critical Theory of Significance.* Champaign: University of Illinois Press, 1989a.

__________. "On Happiness and the Damaged Life." In John O'Neill, ed., *On Critical Theory.* New York: Seabury, 1976a.

__________. "Marcuse and Habermas on New Science." *Polity* 14, no. 2 (Winter 1976b): 158–81.

__________. "Marxism, Feminism, Deconstruction." In Christine Gailey and Viana Muller, eds., *Dialectical Anthropology: The Critical Ethnology of Stanley Diamond.* Forthcoming.

__________. "On Science as Domination." In Alkis Kontos, ed., *Domination.* Toronto: University of Toronto Press, 1975.

__________. *Socio(onto)logy: A Disciplinary Reading.* Champaign: University of Illinois Press, 1989b.

Agger, Ben, and Allan Rachlin. "Left-Wing Scholarship: Current Contradictions of Academic Production." *Humanities in Society* 6, nos. 2–3 (1983): 241–56.

Amsler, Mark, ed. *The Languages of Creativity: Models, Problem-Solving, Discourse.* Newark: University of Delaware Press, 1986.

Anderson, D.C. "Some Organisational Features in the Local Production of a Plausible Text." *Philosophy of the Social Sciences* 8 (1978): 113–35.

Anthony, L.J., H. East, and M. Slater. "The Growth of Literature in Physics." *Reports on Progress in Physics* 32 (1969): 709–67.

Austin, J. L. *How to Do Things with Words.* Oxford: Oxford University Press, 1962.

Bakanic, Von, Clark McPhail, and Rita J. Simon. "The Manuscript Review and Decision-Making Process." *American Sociological Review* 52 (October 1987: 631–42.

Barthes, Roland. "The Death of the Author." In Stephen Heath, ed., *Image-Music-Text: Roland Barthes.* London: Fontana, 1977.

_________. *The Pleasure of the Text.* New York: Hill and Wang, 1974.

_________. *Writing Degree Zero.* New York: Hill and Wang, 1967.

Baudrillard, Jean. *The Mirror of Production.* St. Louis: Telos Press, 1975.

Bazerman, Charles. "What Written Knowledge Does: Three Examples of Academic Discourse." *Philosophy of the Social Sciences* 11 (1981): 361–87.

Becker, Howard. *Writing for Social Scientists.* Chicago: University of Chicago Press, 1986.

Benhabib, Selya. *Critique, Norm and Utopia.* New York: Columbia University Press, 1987.

Benjamin, Walter. "Art in the Age of Mechanical Reproduction." In Benjamin, *Illuminations.* New York: Schocken, 1969.

_________. "The Author as Producer." In Andrew Arato and Eike Gebhardt, eds., *The Essential Frankfurt School Reader.* New York: Urizen, 1978.

Berghe, Pierre L. van den. "Dialectic and Functionalism: Toward Reconciliation." *American Sociological Review* 28 (1963): 695–705.

Bernstein, Basil. *Class, Codes and Control.* London: Routledge and Kegan Paul, 1971.

Bhaskar, Roy. *A Realist Theory of Science.* Sussex: Harvester, 1978.

Birnbaum, M.H. "The Devil Rides Again: Correlation as an Index of Fit." *Psychological Bulletin* 79 (1973): 239–42.

Black, Max. *Models and Metaphors.* Ithaca: Cornell University Press, 1962.

Blackburn, Robin, ed. *Ideology in Social Science.* London: Fontana, 1972.

Bloom, Allan. *The Closing of the American Mind.* New York: Simon and Schuster, 1987.

Bloom, Harold. *A Map of Misreading.* New York: Oxford University Press, 1975.

Blum, Alan. *Theorizing.* London: Heinemann, 1975.

Bourdieu, Pierre. "The Specificity of the Scientific Field and the Social Conditions of the Progress of Reason." *Social Science Information* 14, 6 (1975): 47.

Bradshaw, York W. "Urbanization and Underdevelopment: A Global Study of Modernization, Urban Bias, and Economic Dependency." *American Sociological Review* 52 (April 1987): 224–39.

Brodkey, Linda. *Academic Writing as Social Practice.* Philadelphia: Temple University Press, 1987.

Brown, Courtney. "Voter Mobilization and Party Competition in a Volatile Electorate." *American Sociological Review* 52 (February 1987): 59–72.

Brustein, William. "The Political Geography of Belgian Fascism: The Case of Rexism." *American Sociological Review* 53 (February 1988): 69–80.

Callinicos, Alex. "Postmodernism, Post-structuralism, Post-Marxism?" *Theory, Culture and Society* 2, no. 3 (1985): 85–101.

Camic, Charles. "The Making of a Method: A Historical Reinterpretation of the Early Parsons." *American Sociological Review* 52 (August 1987): 421–39.

Cicourel, Aaron. *Method and Measurement in Sociology.* Glencoe, Ill.: Free Press, 1964.

Cohen, Jere, Lawrence Hazelrigg, and Whitney Pope. "DeParsonizing Weber: A Critique of Parsons' Interpretation of Weber's Sociology." *American Sociological Review* 40 (1975): 229–41.

Cohen, Lawrence E., and Kenneth C. Land. "Age Structure and Crime: Symmetry Versus Asymmetry and the Projection of Crime Rates Through the 1990s." *American Sociological Review* 52 (April 1987): 170–83.

Cohen, Walter, and Peter Hohendahl. "Marxist Literary Critics: Problems and Proposals." *Humanities in Society* 6, nos. 2–3 (Spring-Summer 983): 161–77.

Cole, Stephen, Gary Simon, and Jonathan R. Cole. "Do Journal Rejection Rates Index Concensus?" *American Sociological Review* 53 (February 1988): 152–56.

Coleman, James S. *An Introduction to Mathematical Sociology.* New York: Free Press, 1964.

_________. "Social Theory, Social Research and a Theory of Action." *American Journal of Sociology* 91, no. 6 (1986): 1309–35.

Collins, Randall. "On the Microfoundations of Macrosociology." *American Journal of Sociology* 86 (1981): 984–1014.

Coser, Lewis, Charles Kadushin, and Walter Powell. *Books: The Culture and Commerce of Publishing*. New York: Basic Books, 1982.

Coulthard, Malcolm, and Martin Montgomery, eds. *Studies in Discourse Analysis*. London: Routledge and Kegan Paul, 1981.

Coward, Rosalind, and John Ellis. *Language and Materialism: Developments in Semiology and the Theory of the Subject*. London: Routledge and Kegan Paul, 1977.

Culler, Jonathan. *On Deconstruction*. Ithaca: Cornell University Press, 1982.

Davis, Kingsley, and W.E. Moore. "Some Principles of Stratification." *American Sociological Review* 10 (1942): 242–49.

Day, Robert A. *How to Write and Publish a Scientific Paper*. Philadelphia: ISI Press, 1979.

deMan, Paul. *Allegories of Reading*. New Haven: Yale University Press, 1979.

Derrida, Jacques. *Of Grammatology*. Baltimore: Johns Hopkins University Press, 1976.

__________. *Writing and Difference*. Chicago: University of Chicago Press, 1978.

Diesing, Paul. *Science and Ideology in the Policy Sciences*. New York: Aldine, 1982.

DiMaggio, Paul. "Classification in Art." *American Sociological Review* 52 (August 1987): 440–55.

Dowling, William. *Jameson, Althusser, Marx*. Ithaca: Cornell University Press, 1984.

Durkheim, Emile. *The Rules of Sociological Method*. Glencoe, Ill.: Free Press, 1950.

Dye, James Wayne. "The Poetization of Science." In Mark Amsler, ed., *The Languages of Creativity*. Newark: University of Delaware Press, 1986.

Eagleton, Terry. *Literary Theory*. Minneapolis: University of Minnesota Press, 1983.

__________. *Marxism and Literary Criticism*. London: Methuen, 1976.

__________. "Marxism, Structuralism and Poststructuralism." *Diacritics* 15, no. 4 (Winter 1985): 2–56.

Eco, Umberto. *The Role of the Reader: Explanation in the Semiotics of Texts*. Bloomington: Indiana University Press, 1979.

Fararo, Thomas J. *Mathematical Ideas and Sociological Theory: A Special Issue of the Journal of Mathematical Sociology*. New York: Gordon and Breach, 1984a.

__________. "Neoclassical Theorizing and Formalization in Sociology." *Journal of Mathematical Sociology* 10 (1984b): 361–93.

Faust, David. *The Limits of Scientific Reasoning*. Minneapolis: University of Minneosta Press, 1984.

Feyerabend, Paul. *Against Method*. London: New Left Books, 1975.

Fish, Stanley. *Is There a Text in the Class?* Cambridge: Harvard University Press, 1980.

Fleck, L. *Genesis and Development of a Scientific Fact*. Chicago: University of Chicago Press, 1979.

Foucault, Michel. *The Archaeology of Knowledge*. New York: Harper & Row, 1976.

__________. *Discipline and Punish: The Birth of the Prison*. London: Allen Lane, 1979.

Fraser, Nancy. "The French Derrideans: Politicizing Deconstruction or Deconstructing the Political?" *New German Critique*, no. 33 (1984): 127–54.

Friedrichs, Robert. *A Sociology of Sociology*. New York: Free Press, 1970.

Gardner, John. *On Becoming a Novelist*. New York: Harper & Row, 1983.

Garfinkel, Harold. *Studies in Ethnomethodology*. Englewood Cliffs, N.J.: Prentice-Hall, 1967.

Giddens, Anthony. *New Rules of Sociological Method: A Positive Critique of Interpretive Sociologies*. New York: Basic Books, 1976.

Gilbert, G. Nigel. "Referencing as Persuasion." *Social Studies of Science* 7 (1977): 113–22.

__________. "The Transformation of Research Findings into Scientific Knowledge." *Social Studies of Science* 6 (1976): 281–306.

Gilbert, G. Nigel, and Michael Mulkay. *Opening Pandora's Box: A Sociological Analysis of Scientists' Discourse.* Cambridge: Cambridge University Press, 1984.

_________. "Scientists' Theory Talk." *Canadian Journal of Sociology* 8 (1983): 179–97.

Glaser, Barney, and Anselm Strauss. *The Discovery of Grounded Theory.* Chicago: Aldine, 1967.

Gottdiener, M. "Hegemony and Mass Culture: A Semiotic Approach." *American Journal of Sociology* 90, no. 5 (1985): 979–1001.

Gouldner, Alvin. *The Coming Crisis of Western Sociology.* New York: Basic Books, 1970.

_________. *The Dialectic of Ideology and Technology.* New York: Seabury, 1976.

_________. *The Future of the Intellectuals and the Rise of the New Class.* New York: Seabury, 1979.

_________. *The Two Marxisms: Contradictions and Anomalies in the Development of Theory.* New York: Seabury, 1980.

Graff, Gerald. *Literature Against Itself: Literary Ideas in Modern Society.* Chicago: University of Chicago Press, 1977.

Grant, Linda, Kathryn B. Ward, and Xue Lan Rong. "Is There an Association Between Gender and Methods in Sociological Research?" *American Sociological Review* 52 (December 1987): 856–62.

Greer, Scott. *The Logic of Social Inquiry.* Chicago: Aldine, 1969.

Gusfield, Joseph. "The Literary Rhetoric of Science." *American Sociological Review* 41 (1976): 16–34.

Habermas, Jurgen. *Communication and the Evolution of Society.* Boston: Beacon Press, 1979.

_________. *Knowledge and Human Interests.* Boston: Beacon Press, 1971.

_________. "Modernity versus Postmodernity." *New German Critique,* no. 22 (1981): 3–14.

_________. "Technology and Science as 'Ideology.'" In Habermas, *Toward a Rational Society.* Boston: Beacon Press, 1970.

_________. *The Theory of Communicative Action.* Vols. 1 and 2. Boston: Beacon Press, 1984, 1987.

Hagan, John, A.R. Gillis, and John Simpson. "The Class Structure of Gender and Delinquency: Toward a Power-Control Theory of Common Delinquent Behavior." *American Journal of Sociology* 90, no. 6 (1985): 1151–75.

Halaby, Charles N. "Worker Attachment and Workplace Authority." *American Sociological Review* 51 (October 1986): 634–49.

Hall, Everett Wesley. *Modern Science and Human Values.* New York: Dell, 1966.

Hallinan, Maureen T., and Richard A. Williams. "The Stability of Students' Interracial Friendships." *American Sociological Review* 52 (October 1987): 653–64.

Harari, Josue V., ed. *Textual Strategies.* Ithaca: Cornell University Press, 1979.

Hargens, Lowell L. "Further Evidence of Field Differences in Consensus from the NSF Peer Review Studies." *American Sociological Review* 53 (February 1988): 157–60.

_________. "Scholarly Consensus and Journal Rejection Rates." *American Sociological Review* 53 (February 1988): 139–51.

Hartman, Geoffrey. *Saving the Text: Literature, Derrida, Philosophy.* Baltimore: Johns Hopkins University Press, 1981.

Hayek, Friedrich A. *The Road to Serfdom.* Chicago: University of Chicago Press, 1944.

Heisenberg, Werner. *Philosophical Problems of Quantum Physics.* Woodbridge, Conn.: Ox Bow Press, 1979.

Hempel, Carl G. *Philosophy of Natural Science.* Englewood Cliffs, N.J.: Prentice-Hall, 1966.

Hesse, Mary. *Revolutions and Reconstructions in the Philosophy of Science.* Bloomington: Indiana University Press, 1980.

Hirsch, E.D. *Validity in Interpretation.* New Haven: Yale University Press, 1967.

Hodgkin, Luke. "Mathematics as Ideology and Politics." In Les Levidow, ed., *Radical Science Essays*. London: Free Association, 1986.

Horkheimer, Max. *Critical Theory*. New York: Herder and Herder, 1972.

———. *Eclipse of Reason*. New York: Seabury, 1974.

Horkheimer, Max, and Theodor W. Adorno. *Dialectic of Enlightenment*. New York: Herder and Herder, 1972.

Hout, Michael, and Andrew M. Greeley. "The Center Doesn't Hold: Church Attendance in the United States, 1940–1984." *American Sociological Review* 52 (June 1987): 325–45.

Husserl, Edmund. *The Crisis of the European Sciences and Transcendental Phenomenology*. Evanston: Northwestern University Press, 1970.

Huyssen, Andreas. "Mapping the Postmodern." *New German Critique* 33 (1984): 5–52.

Institute for Social Research. *Aspects of Sociology*. Boston: Beacon Press, 1972.

Iser, Wolfgang. *The Act of Reading: A Theory of Aesthetic Response*. Baltimore: Johns Hopkins University Press, 1978.

———. *The Implied Reader: Patterns of Communication in Prose Fiction from Bunyan to Beckett*. Baltimore: Johns Hopkins University Press, 1974.

Jacoby, Russell. "A Falling Rate of Intelligence?" *Telos 27* (1976): 141–46.

———. *The Last Intellectuals: American Culture in the Age of Academe*. New York: Basic Books, 1987.

Jameson, Fredric. "Ideology of the Text." *Salmagundi* 31 (1975–76): 204–46.

———. *The Political Unconscious: Narrative as a Socially Symbolic Act*. Ithaca: Cornell University Press, 1981.

———. "Postmodernism, or the Cultural Logic of Late Capitalism." *New Left Review* 146 (July-August 1984): 53–92.

———. *The Prison-House of Language*. Princeton: Princeton University Press, 1972.

———. "Science Versus Ideology." *Humanities in Society* 6, nos. 2–3 (1983): 283–302.

Jay, Martin. *Marxism and Totality*. Berkeley: University of California Press, 1984.

Jenkins, J. Craig, and Craig M. Eckert. "Channeling Black Insurgency: Elite Patronage and Professional Social Movement Organizations in the Development of the Black Movement." *American Sociological Review* 51 (December 1986): 812–29.

Kahn, Joan R., and William M. Mason. "Political Alienation, Cohort Size, and the Easterlin Hypothesis." *American Sociological Review* 52 (April 1987): 155–69.

Kaplan, Abraham. The *Conduct of Inquiry*. San Francisco: Chandler, 1964.

Kaufman, Robert L., and Paul G. Schervish. "Using Adjusted Crosstabulations to Interpret Log-Linear Relationships." *American Sociological Review* 51 (October 1986): 717–33.

Keller, Evelyn Fox. *Reflections on Gender and Science*. New Haven: Yale University Press, 1985.

Kingston, Paul William, and Steven L. Nock. "Time Together Among Dual-Earner Couples." *American Sociological Review* 52 (June 1987): 391–400.

Kiser, Edgar, and Kriss A. Drass. "Changes in the Core of the World-System and the Production of Utopian Literature in Great Britain and the United States, 1883–1975." *American Sociological Review* 52 (April 1987): 286–93.

Knorr-Cetina, Karin D. *The Manufacture of Knowledge: An Essay on the Constructivist and Contextual Nature of Science*. New York: Pergamon, 1981.

———. "Producing and Reproducing Knowledge: Descriptive or Constructive? Toward a Model of Research Production." *Social Science Information* 16 (1977): 969–96.

———. "Tinkering Toward Success: Prelude to a Theory of Scientific Practice." *Theory and Society* 8 (1979): 347–76.

Knorr-Cetina, Karin D., and Aaron Cicourel, eds. *Advances in Social Theory and Methodology: Toward an Integration of Micro- and Macro-Sociologies*. London: Routledge and Kegan Paul, 1981.

Knorr-Cetina, Karin D., and D. Knorr. *From Scenes to Scripts: On the Relationship Between Research and Publication in Science.* Research Monograph No. 132. Vienna: Institute for Advanced Studies, 1978.

Knorr-Cetina, Karin D., and Michael Mulkay, eds. *Science Observed: Perspectives on the Social Study of Science.* Beverly Hills: Sage, 1983.

Kristeva, Julia. *Desire in Language.* New York: Columbia University Press, 1980.

Lakatos, Imre, and Alan Musgrove, eds. *Criticism and the Growth of Knowledge.* Cambridge: Cambridge University Press, 1970.

Land, Kenneth, C. "Mathematical Formalization of Durkheim's Theory of Division of Labor." In E. Borgatta and G. Bohrnstedt, eds., *Sociological Methodology.* San Francisco: Jossey-Bass, 1970.

Latour, Bruno, and Steve Woolgar. *Laboratory Life: The Social Construction of Scientific Facts.* Beverly Hills: Sage, 1979.

Lave, C.A., and J.G. March. *An Introduction to Models in the Social Sciences.* New York: Harper & Row, 1975.

Law, J., and R. Williams. "Putting Facts Together: A Study of Scientific Persuasion." *Social Studies of Science* 12, no. 4 (1982): 535–58.

Lentricchia, Frank. *After the New Criticism.* Chicago: University of Chicago Press, 1980.

Levidow, Les., ed. *Radical Science Essays.* London: Free Association, 1986.

Levi-Strauss, Claude. *The Savage Mind.* Chicago: University of Chicago Press, 1966.

Link, Bruce G. "Understanding Labeling Effects in the Area of Mental Disorders: An Assessment of the Effects of Expectations of Rejection." *American Sociological Review* 52 (February 1987): 96–112.

Lodahl, Janice B., and Gerald Gordon. "The Structure of Scientific Fields and the Funding of University Graduate Departments." *American Sociological Review* 37 (1972): 57–72.

London, Bruce. "Structural Determinants of Third World Urban Change: An Ecological and Political Economic Analysis." *American Sociological Review* 52 (February 1987): 28–43.

Lukacs, Georg. *History and Class Consciousness.* London: Merlin, 1971.

Luke, Timothy W. *Screens of Power: Ideology, Domination and Resistance in Informational Society.* Champaign: University of Illinois Press, 1989.

Lynch, Michael. *Art and Artifact in Laboratory Science.* London: Routledge and Kegan Paul, 1981.

Maniquis, Robert M., ed. "Marxists and the University." *Humanities in Society* 6, nos. 2–3 (Spring-Summer 1983).

Marcuse, Herbert. *An Essay on Liberation.* Boston: Beacon Press, 1969.

__________. *One-Dimensional Man.* Boston: Beacon Press, 1964.

Mare, Robert D., and Meichu D. Chen. "Further Evidence on Sibship Size and Educational Stratification." *American Sociological Review* 51 (June 1986): 403–12.

Martindale, Donald. *The Nature and Types of Sociological Theory.* 2d ed. Boston: Houghton Mifflin, 1981.

Marx, Karl, and Friedrich Engels. *The German Ideology.* New York: International Publishers, 1949.

Masterman, Margaret. "The Nature of a Paradigm." In Imre Lakatos and Alain Musgrove, eds., *Criticism and the Growth of Knowledge.* Cambridge: Cambridge University Press, 1970.

McHoul, A.W. *Telling How Texts Talk.* London: Routledge and Kegan Paul, 1982.

McHugh, Peter, et al. *On the Beginning of Social Inquiry.* London: Routledge and Kegan Paul, 1974.

Medvedev, Pavel Nikolaevich. *The Formal Method in Literary Scholarship: A Critical Introduction to Sociological Poetics.* Baltimore: Johns Hopkins University Press, 1978.

Mehan, Hugh, and Houston Wood. *The Reality of Ethnomethodology,* New York: Wiley, 1975.

Mendelsohn, Everett, Peter Weingart, and Richard Whitley. *The Social Production of Scientific Knowledge.* Boston: D. Reidel, 1977.

Merton, Robert K. "Paradigm for the Sociology of Knowledge." In Norman W. Storer, ed., *The Sociology of Science: Theoretical and Empirical Investigations.* Chicago: University of Chicago Press, 1973.

_________. "Priorities in Scientific Discovery: A Chapter in the Sociology of Science." *American Sociological Review* 22 (December 1957): 655–59.

_________. "On Sociological Theories of the Middle Range." In Merton, *On Theoretical Sociology.* New York: Free Press, 1967.

_________. *The Sociology of Science.* Chicago: University of Chicago Press, 1973.

Messner, Steven F., and Scott J. South. "Economic Deprivation, Opportunity Structure, and Robbery Victimization: Intra- and Interracial Patterns." *Social Forces* 64, no. 4 (1986): 975–91.

Miller, Karen A., Melvin L. Kohn, and Carmi Schooler. "Educational Self-Direction and Personality." *American Sociological Review* 51 (June 1986): 372–90.

Mokrzycki, Edmund. *Philosophy of Science and Sociology.* London: Routledge and Kegan Paul, 1983.

Mueller, Claus. *The Politics of Communication.* New York: Oxford University Press, 1973.

Mulkay, Michael. "Action and Belief or Scientific Discourse? A Possible Way of Ending Intellectual Vassalage in Social Studies of Science." *Philosophy of the Social Sciences* 11 (1981): 163–71.

_________. *Science and the Sociology of Knowledge.* London: Allen and Unwin, 1979.

_________. *The Word and the World: Explorations in the Form of Sociological Analysis.* Winchester: Allen and Unwin, 1985.

Mulkay, Michael, and G. Nigel Gilbert. "Accounting for Error: How Scientists Construct Their Social World When They Account for Correct and Incorrect Belief." *Sociology* 16 (1982): 165–83.

_________. "What is the Ultimate Question? Some Remarks in Defense of the Analysis of Scientific Discourse." *Social Studies of Science* 12, no. 2 (1982): 309–19.

Mulkay, Michael, Jonathan Potter, and Steven Yearley. "Why an Analysis of Scientists' Discourse is Needed." In Karin D. Knorr-Cetina and Michael Mulkay, eds., *Science Observed: Perspectives on the Social Study of Science.* Beverly Hills: Sage, 1983.

Mullins, Nicholas. *Theories and Theory Groups in Contemporary American Sociology.* New York: Harper & Row, 1973.

Mullins, N., L. Hargens, P.K. Hecht, and E.L. Kick. "The Group Structure of Co-Citation Clusters: A Comparative Study." *American Sociological Review* 42 (1977): 552–62.

Nagel, Ernest. *The Structure of Science.* New York: Harcourt, Brace and World, 1961.

Nelson, Cary, and Lawrence Grossman., eds. *Marxism and the Interpretation of Culture.* Champaign: University of Illinois Press, 1988.

Nowak, Leszek. *The Structure of Idealization: Towards a Systematic Interpretation of the Marxian Idea of Science.* Boston: D. Reidel, 1980.

O'Kane, John. "Marxism, Deconstruction and Ideology: Notes towards an Articulation." *New German Critique,* no. 33 (1984): 219–47.

O'Neill, John. "Decolonization and the Ideal Speech Community: Some Issues in the Theory and Practice of Communicative Competence." In John Forester, ed., *Critical Theory and Public Life.* Cambridge: MIT Press, 1985.

_________. *Essaying Montaigne: A Study of the Renaissance Institution of Writing and Reading.* London: Routledge and Kegan Paul, 1982.

_________. "The Literary Production of Natural and Social Science Inquiry." *Canadian Journal of Sociology* 6 (1981): 105–20.

_________. "Marxism and the Two Sciences." *Philosophy of the Social Sciences* 11 (1981): 281–302.

Paci, Enzo. *The Function of the Sciences and the Meaning of Man.* Evanston: Northwestern University Press, 1972.

Palmer, Donald, Roger Friedland, and Jitendra V. Singh. "The Ties That Bind: Organizational and Class Bases of Stability in a Corporate Interlock Network." *American Sociological Review* 51 (Decmeber 1986): 781–96.

Palmer, Richard. *Hermeneutics.* Evanston: Northwestern University Press, 1969.

Parker, Andrew. "Of Politics and Limits: Derrida Re-Marx." *SCE Reports* 8 (1980): 83–104.

Pett, Joseph C. *Pictures, Images, and Conceptual Change.* Boston: D. Reidel, 1981.

Phillips, Derek. *Knowledge From What? Theories and Methods in Social Research.* Chicago: Rand McNally, 1971.

_________. "Paradigms and Incommensurability." *Theory and Society* 2 (1975): 37–62.

Pope, Whitney. "Classic on Classic: Parsons' Interpretation of Durkheim." *American Sociological Review* 38 (1973): 399–415.

Pope, Whitney, Jere Cohen, and Lawrence E. Hazelrigg. "On the Divergence of Weber and Durkheim: A Critique of Parsons' Convergence Thesis." *American Sociological Review* 40 (1975): 417–27.

Popper, Karl. *The Logic of Scientific Discovery.* New York: Science Edition, 1961.

Potter, Jonathan, and Michael Mulkay. "Making Theory Useful." *Fundamenta Scientiae* 3 (1983): 259–78.

Potter, Jonathan, P. Stringer, and M. Wetherell. *Social Texts and Contexts: Literature and Social Psychology.* London: Routledge and Kegan Paul, 1983.

Rabinowitz, Peter. "Truth in Fiction: A Reexamination of Audiences." *Critical Inquiry* 4 (1977): 121–42.

Rapoport, Anatol. *Mathematical Models in the Social and Behavioral Sciences.* New York: Wiley, 1983.

Rasmussen, David. "Communicative Action and the Fate of Modernity." *Theory, Culture and Society* 2, no. 3 (1985): 133–44.

Restivo, S. "Some Perspectives in Contemporary Sociology of Science." *Science, Technology and Human Values* 35 (1981): 22–30.

Reynolds, Larry, and Janice M. Reynolds., eds. *The Sociology of Sociology.* New York: McKay, 1970.

Ricoeur, Paul. *Hermeneutics and the Human Sciences.* Cambridge: Cambridge University Press, 1981.

_________. "The Model of the Text: Meaningful Action Considered as a Text." *Social Research* 38 (1970): 529–62.

Riffaterre, Michael. *Semiotics of Poetry.* Bloomington: Indiana University Press, 1978.

Ritzer, George. *Sociology: A Multiple Paradigm Science.* Rev. ed. Boston: Allyn and Bacon, 1980.

_________. *Toward an Integrated Sociological Paradigm: The Search for an Exemplar and an Image of the Subject Matter.* Boston: Allyn and Bacon, 1981.

Robinson, Robert V., and Maurice A. Garnier. "Class Reproduction Among Men and Women in France: Reproduction Theory on Its Home Ground." *American Journal of Sociology* 91, no. 2 (1985): 250–80.

Rorty, Richard. "Philosophy as a Kind of Writing: An Essay on Derrida." *New Literary History* 10 (1978): 141–60.

Rose, Hilary, and Steven Rose., eds. *The Radicalization of Science: Ideology of/in the Natural Sciences.* London: Macmillan, 1976.

Ryan, Michael. *Marxism and Deconstruction.* Baltimore: Johns Hopkins University Press, 1982.

Sartre, Jean-Paul. *What Is Literature?* New York: Harper & Row, 1965.

Saussure, Ferdinand de. *Course in General Linguistics.* London: Peter Owen, 1960.

Schmidt, James. "Reification and Recollection: Emancipatory Interests and the Sociology of Knowledge." *Canadian Journal of Political and Social Theory* 2, no. 1 (Winter 1978): 89–111.

Sciulli, David. "Voluntaristic Action as a Distinct Concept: Theoretical Foundations of Societal Constitutionalism." *American Sociological Review* 51 (December 1986): 743–66.

Searle, John. *Speech Acts.* Cambridge: Cambridge University Press, 1969.

Sewart, John J. "Critical Theory and the Critique of Conservative Method." *American Sociologist* 13 (1978): 15–22.

Sharratt, B. *Reading Relations: Structures of Literary Production.* Brighton: Harvester, 1982.

Silverman, D., and B. Torode. *The Material Word.* London: Routledge and Kegan Paul, 1980.

Simpson, Sally S. "The Decomposition of Antitrust: Testing a Multi-Level, Longitudinal Model of Profit-Squeeze." *American Sociological Review* 51 (December 1986): 859–75.

Smith-Lovin, Lynn, John V. Skvoretz, and Charlotte G. Hudson. "Status and Participation in Six-Person Groups: A Test of Skvoretz's Comparative Status Model." *Social Forces* 64, no. 4 (1986): 992–1005.

Sobel, Michael E., Michael Hout, and Otis Dudley Duncan. "Exchange, Structure, and Symmetry in Occupational Mobility." *American Journal of Sociology* 91, no. 2 (1985): 359–72.

Sohn-Rethel, Albert. "Science as Alienated Consciousness." *Radical Science Journal* 2/3 (1975): 72–101.

Sontag, Susan. *On Photography.* New York: Farrar, Straus and Giroux, 1977.

Stahura, John M. "Suburban Socioeconomic Status Change: A Comparison of Models, 1950–1980." *American Sociological Review* 52 (April 1987): 268–77.

Stockman, Norman. *Antipositivist Theories of the Sciences.* Dordrecht: D. Reidel, 1984.

Struik, Dirk J. "The Sociology of Mathematics." *Science and Society* 6 (1942): 58–70.

Sullivan, D., H.D. White, and E.J. Barboni. "Co-Citation Analysis of Science: An Evaluation." *Social Studies of Science* 7 (1977): 223–40.

Sutton, John R. "Doing Time: Dynamics of Imprisonment in the Reformist State." *American Sociological Review* 52 (October 1987): 612–30.

Thomas, Melvin E., and Michael Hughes. "The Continuing Significance of Race: A Study of Race, Class, and Quality of Life in America, 1972–1985." *American Sociological Review* 51 (December 1986): 830–41.

Thompson, John B. *Studies in the Theory of Ideology.* Berkeley: University of California Press, 1984.

Tienda, Marta, Shelley A. Smith, and Vilma Ortiz. "Industrial Restructuring, Gender Segregation, and Sex Differences in Earnings." *American Sociological Review* 52 (April 1987): 195–210.

Turner, Jonathan H. "Toward a Sociological Theory of Motivation." *American Sociological Review* 52 (February 1987): 15–27.

Udry, J. Richard, and John O.G. Billy. "Initiation of Coitus in Early Adolescence." *American Sociological Review* 52 (December 1987): 841–55.

Van den Berg, Axel. "Critical Theory: Is There Still Hope?" *American Journal of Sociology* 86, no. 3 (1980): 449–78.

Weil, Frederick D. "Cohorts, Regimes, and the Legitimation of Democracy: West Germany Since 1945." *American Sociological Review* 52 (June 1987): 308–24.

Wellmer, Albrecht W. "Communications and Emancipation: Reflections on the Linguistic Turn in Critical Theory." In John O'Neill, ed., *On Critical Theory.* New York: Seabury, 1976.

Wells, Susan. "Jurgen Habermas, Communicative Competence, and the Teaching of Technical Discourse." In Cary Nelson, ed., *Theory in the Classroom.* Champaign: University of Illinois Press, 1986.

Wharton, Amy S., and James N. Baron. "So Happy Together? The Impact of Gender Segregation on Men at Work." *American Sociological Review* 52 (October 1987): 574–87.

Wilner, Patricia. "The Main Drift of Sociology Between 1936 and 1984." *History of Sociology* 5, no. 2 (1985): 1–20.

Wilson, T. "On the Role of Mathematics in the Social Sciences." *Journal of Mathematical Sociology* 10 (1984): 221–39.

Winch, Peter. *The Idea of a Social Science.* London: Routledge and Kegan Paul, 1969.

Wittgenstein, Ludwig. *Philosophical Investigations.* Oxford: Basil Blackwell, 1953.

_________. *Remarks on the Foundations of Mathematics.* Cambridge: MIT Press, 1978.

Woolgar, S. "Discovery: Logic and Sequence in a Scientific Text." In K. Knorr, R. Krohn, and R. Whitley, eds., *The Social Process of Scientific Investigation. Sociology of the Sciences.* Yearbook Vol. 4. Dordrecht: D. Reidel, 1980.

Yearley, Steven. *Science and Sociological Practice.* London: Open University Press, 1984.

_________. "Textual Persuasion: The Role of Social Accounting in the Construction of Scientific Arguments." *Philosophy of the Social Sciences* 11 (1981): 409–35.

Zetterberg, Hans. *On Theory and Verification in Sociology.* Totowa, N.J.: Bedminister, 1955.

Zito, George V. *Systems of Discourse: Structure and Semiotics in the Social Sciences.* New York: Greenwood, 1986.

Index